Fuzziness in Information Systems

Miroslav Hudec

Fuzziness in Information Systems

How to Deal with Crisp and Fuzzy Data
in Selection, Classification,
and Summarization

 Springer

Miroslav Hudec
Faculty of Economic Informatics
University of Economics in Bratislava
Bratislava
Slovakia

ISBN 978-3-319-82598-4   ISBN 978-3-319-42518-4   (eBook)
DOI 10.1007/978-3-319-42518-4

© Springer International Publishing Switzerland 2016
Softcover reprint of the hardcover 1st edition 2016
This work is subject to copyright. All rights are reserved by the Publisher, whether the whole or part
of the material is concerned, specifically the rights of translation, reprinting, reuse of illustrations,
recitation, broadcasting, reproduction on microfilms or in any other physical way, and transmission
or information storage and retrieval, electronic adaptation, computer software, or by similar or dissimilar
methodology now known or hereafter developed.
The use of general descriptive names, registered names, trademarks, service marks, etc. in this
publication does not imply, even in the absence of a specific statement, that such names are exempt from
the relevant protective laws and regulations and therefore free for general use.
The publisher, the authors and the editors are safe to assume that the advice and information in this
book are believed to be true and accurate at the date of publication. Neither the publisher nor the
authors or the editors give a warranty, express or implied, with respect to the material contained herein or
for any errors or omissions that may have been made.

Printed on acid-free paper

This Springer imprint is published by Springer Nature
The registered company is Springer International Publishing AG Switzerland

I dedicate this book to Martina. I would like to skip writing reasons, because this part should be shorter than a book chapter.

Foreword

It is a pleasure to write this foreword because this book is an important contribution to the literature on applications of fuzzy models. There are many books dealing with fuzzy sets in a general way but this work is an essential contribution to the description of fuzziness in information systems.

Usually in statistical information systems data are stored as numbers which pretend a precision which is not justified, because real data are frequently not available as precise numbers but they are more or less non-precise. This imprecision is different from errors and it is best modelled by the so-called fuzzy numbers, which are special fuzzy subsets of the set of real numbers. To describe fuzziness in quantitative mathematical terms, is an important innovation in science and management.

When Karl Menger introduced the generalization of classical sets in the year 1951 by generalizing the indicator function of classical sets, this was a theoretical concept and it took many years until practical applications of these generalized sets came up. An important step was the paper by Lotfi Zadeh in 1965 when he introduced the name *fuzzy set* and defined generalized set operations based on the defining functions of fuzzy sets. These defining functions were called membership functions and in the last decades of the twentieth century enormous research activities developed the theory and applications of fuzzy sets. The concept of fuzziness was extended to generalize logics, and the calculus of fuzzy logics was created. In the meantime fuzzy concepts are applied in many scientific fields, for example in civil engineering for risk analysis of structures, in medical science for diagnostic systems, in measurement science to describe results of precision measurement, in statistics for the description and analysis of real data, in Bayesian analysis to model fuzzy a priori information, in information science to describe fuzzy information and to formulate fuzzy questions.

Chapter 1 of the book gives an introduction to fuzzy sets and fuzzy logic, linguistic variables, fuzzy quantifiers, and related references.

Chapter 2, Fuzzy Queries, considers the way from crisp to fuzzy queries, the construction of fuzzy sets for flexible conditions, the conversion of fuzzy conditions to SQL ones, the calculation of matching degrees, empty and overabundant answers, and some issues related to practical realization.

Chapter 3, Linguistic Summaries, explains the benefits of linguistic summarization (LS), the basic structure of LS, relative quantifiers, quality measures of LS, applicability of LS, and building summaries.

Chapter 4, Fuzzy Inference, is devoted to fuzzy models for control systems. After introducing fuzzy inference engines, the chapter presents a section on fuzzy classification, and concludes with remarks to applications.

The next chapter, Fuzzy Data in Relational Databases, is central to the book. It starts with the classical relational databases, has a section on fuzziness in the real world, explains fuzzy databases and their basic model, fuzzy data in traditional relational databases, aggregation functions in queries, and linguistic summaries on fuzzy data.

Chapter 6, Perspectives, Synergies and Conclusion, briefly explains the relationship between fuzzy inference and fuzzy databases and linguistic summaries as well as fuzzy queries.

The references at the end of each chapter are helpful for further reading.

Appendixes, Illustrative Interfaces and Applications for Fuzzy Queries and Illustrative Interfaces and Applications for Linguistic Summaries, and an Index are helpful for the reader.

This book explains important applications of fuzzy logic in information systems. Congratulations to the author for this valuable and up-to-date contribution to information science.

Wien   Reinhard Viertl
April 2016

Preface

The increasing use of information systems by governmental agencies and businesses has created mountains of data that contain potentially valuable knowledge. Admittedly, these data do constitute "golden mines" which should be swiftly and efficiently processed and interpreted to be useful. Users (e.g. decision-makers) would like to efficiently reveal relevant data. Moreover, users are often not interested in large sheets of figures, but in knowledge that is usually overshadowed by large amount of data.

People can relatively easily answer imprecise questions like, *is it true that most of tall persons in the room wear blue or green shirts*? Different hues of these colours as well as the meaning of the vague term *tall people* are not limitations for solving this task. However, if we want to know, which of these two sentences: *most of young commuters commute short distances; most of medium aged commuters commute short distances* better explains the commuting behaviour, then we have to adapt this query to mine the truth value form the data. The same holds for querying *cheap hotel with good references and if possible near to the city centre* and common-sense reasoning: *if customer buys products very often, then provide high discount.*

The initial research in the theory of fuzzy sets and fuzzy logic was motivated by the perception that traditional computing techniques are not effective in dealing with problems, in which vagueness, imprecision and subjectivity are immanent, and therefore should not be neglected. These types of uncertainty are commonly called fuzziness.

According to Prof. Zadeh, four principal rationales for handling fuzziness exist. Two of them, which are relevant for this book, are: "don't need rationale" and "don't know rationale". In the former, the tolerance for imprecision is in accord with the remarkable human capability to solve variety of tasks without precise calculations. For example, summarizing data by short questions of natural language; creating queries with flexible conditions and approximate inference. In the latter, the values of attributes are not known with sufficient precision to justify the use of traditional databases for storing these data. Many data cannot be adequately expressed as precise numbers or as one linguistic term, due to non-sharp boundaries of observations, tendency of people to estimate or guess answers in surveys and tolerance intervals of measurement instruments. Therefore, the data are often vague and include both quantitative and qualitative elements. Storing these data as crisp values might cause loss of valuable information.

Keeping in mind the aforementioned facts, fuzzy queries, fuzzy inference processes, linguistic summaries and managing fuzzy data in information systems could be the option. We have chosen these areas, because businesses of all sizes and governmental agencies cope with them in their work. The motivation for this book has arisen from the author's experience in teaching courses of fuzzy logic for business informatics and database design and in research and development of information systems and data mining applications mainly for the official statistics purposes. Furthermore, many small- and medium-sized enterprises cannot afford sophisticated tools or experts for information systems and data mining, even though they are aware of limitation of sharp boundaries in data analyses. Many tasks can be solved in a classical way, but their complexity becomes high. The complexity of the problem can be reduced by including the intensity of the examined property. This permits us to discern elements with the same property, based on the intensity matching it.

Roughly, the intent of the book could be depicted in Fig. 1. The usual scenario is that user wants to retrieve data or summarized information from a database. Furthermore, user might be interested to classify data. Often user is not aware of the nature of collected data or cannot determine sharp criteria. In addition, all data including vague ones are usually stored as crisp values.

In the book we examine these approaches theoretically as well as on the municipal statistics data. The latter is illustrated in appendixes. These data are suitable source, due to larger number of municipalities, which are often very similar in several attributes. Second reason is that some of attributes are fuzzy in their nature, but are limited to crisp values.

We should not expect that domain experts are familiar with the fuzzy logic theory. Therefore, the book demonstrates developing user-friendly interfaces to allow users exploring advantages of fuzzy logic in their tasks. Furthermore,

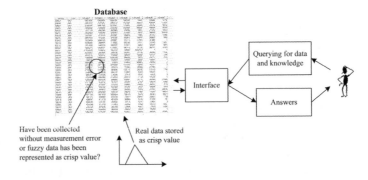

Fig. 1 Intent of the book

companies usually keep data in relational databases. We should keep this in mind during adapting database to cope with the fuzzy data.

The main target audience of the book are students, researchers and practitioners working in fields of data analysis, database design and business intelligence. This book does not go deeply into the foundation and the mathematical theory of fuzzy logic and relational algebra (e.g. theorems' proofs). Hence, intermediate knowledge of fuzzy logic and relational databases is recommended.

The book is divided into six chapters in the following way. Chapter 1 is focused on the theory of fuzzy sets and fuzzy logic to a level, which is advisable to know in order to proceed to next chapters. Readers skilled in fuzzy logic theory can skip this chapter.

Chapter 2 is devoted to flexible queries. The following aspects of flexible queries are examined: constructing fuzzy sets for query conditions; aggregation operators for commutative and non-commutative conditions with and without priorities; dealing with empty and overabundant answer problems and issues related to practical realizations.

Chapter 3 is dedicated to linguistic summaries. We start with the basic linguistic summary and build more complex ones. To meet this goal, selecting appropriate aggregations, implications for preferences and issues related to construction of membership functions are examined. Quality measures of created summaries are also considered. Finally, several possible applicabilities are discussed.

Chapter 4 presents fuzzy logic control architecture adjusted to the aims of business and governmental agencies. It shows fuzzy rules, construction of fuzzy sets and procedures for solving inference tasks by generalized *modus ponens*. In the first part we explain reasoning procedures. In the second part fuzzy expert systems are discussed. In the last part classification by IF-THEN rules is examined.

Chapter 5 covers fuzzification of classical relational databases. We briefly review classical relational databases and fuzzy database models. The emphasis is on storing fuzzy data in classical relational databases in a way that existing data and normal forms are not affected. Furthermore, practical aspects of user-friendly interfaces for storing, updating, querying and summarizing are examined.

Chapter 6 shortly discusses possible integration of fuzzy queries, summarization and inference related to crisp and fuzzy databases. Use of these approaches in a complementary, rather than competitive way, can support variety of tasks.

Finally, we suppose that the book will provoke at least some interest to continue research and also will be of support for developing tailored applications communicating with users by easy-to-use interfaces. Maybe the next generations of relational database management systems and applications will include many fuzzy characteristics and users will enjoy easy-to-use interfaces for fuzzy queries, fuzzy inferences, fuzzy summarization, fuzzy recommending and so on, without the necessity of knowing mathematics of fuzzy logic. We hope that the book will contribute to this field with a membership degree greater than 0.25.

Bratislava
April 2016

Miroslav Hudec

Acknowledgements

Advice and suggestions touching theoretical as well as practical implications related to my work were valuable source for this research. In this direction, I express my deep appreciation to Tomáš Bacigál, Simona Balbi, Mojca Bavdaž, Piet Daas, Boris Delibašić, Erich Peter Klement, Gabriela Kristová, Andreas Meier, Radko Mesiar, Dušan Praženka, Dragan Radojević, Agnieszka Stawinoga, František Sudzina, Matthias Templ, Valentin Todorov, Vanessa Torres van Grinsven, Steven Vale, Enrique Herrera-Viedma, Reinhard Viertl, Miljan Vučetić and Mirko Vujošević.

I also address my gratitude to the authors of publications listed in the references. Their contributions were inspiration for my work and are reflected in the book.

Concerning data required for experiments I thank Statistical Office of the Slovak Republic for providing the data. I am grateful to Eleonora Vallová for proofreading and editing my manuscript.

I am also grateful to Springer for offering me opportunity to publish my humble view and opinion about the fuzziness in their publishing house and especially Ralf Gerstner for the cooperation during the whole process.

My final thanks go to the heads of the Faculty of Economic Informatics and especially Department of Applied Informatics for providing me with space for research which enabled me to spend long hours in researching and writing the book.

Contents

Acronyms

1NF	First Normal Form
2NF	Second Normal Form
3NF	Third Normal Form
COG	Center of Gravity
COM	Center of Maxima
CRM	Customer Relationship Management
DBMS	Database Management System
DDL	Data Definition Language
DNF	Disjunctive Normal Form
FIS	Fuzzy Inference System
FMB	Fuzzy Meta-knowledge Base
FMM	Fuzzy Meta Model
GLC	Generalized Logical Condition
GMP	Generalized Modus Ponens
HDM	Height Defuzzification Method
HOF	Half of Field
IBA	Interpolative Realization of Boolean Algebra
LOM	Left of Maxima
LSs	Linguistic Summaries
MOM	Mean of Maxima
NFE	Necessarily Fuzzy Equal to
NFL	Necessarily Fuzzy Less than
OWA	Ordered Weighted Averaging Operator
PFE	Possibly Fuzzy Equal to
PFL	Possibly Fuzzy Less than
RDBMS	Relational Database Management System
ROM	Right of Maxima
SQL	Structured Query Language

Notations

In each field variables, sets, properties, functions, etc. are either marked by letters using an (informal) agreement, or by letters the authors decided to use in their seminal papers. In this textbook we have decided to keep usual notation of most used terms and adjust the notation of other terms in order to avoid misinterpretation. As it is not always possible, some letters are not used for a single term only, but the explanation on their usage avoids misinterpretation of used letters and indexes.

A	Attribute, fuzzy set, answer to query
$\bar{A}$	Complement of fuzzy set
$A^{(\alpha)}$	α-cut of fuzzy set
A_c	Accuracy
a, m, b	Parameters of fuzzy sets
B	Fuzzy set
$\text{core}(A)$	Core of fuzzy set
$\text{card}(A)$	Cardinality of fuzzy set, $\|A\|$
c	Negation
c_s	Standard negation
c_g	Gödel negation
c_{dg}	Dual Gödel negation
C	Coverage
i_C	Coverage index
D	Domain, distance
d	Measure of fuzziness
E_c	Specificity
F	Fuzzification operator, fuzzy set
h	Height of fuzzy set
H	Highest value of attribute in a database
i_{sKD}	Kleene–Dienes implication

(continued)

(continued)

i_{sL}	Łukasiewicz implication
i_{qZ}	Zadeh implication
i_{rGd}	Gödel implication
i_{rGg}	Goguen implication
L	Lowest value of attribute in a database
N	Negative preference
O	Outlier measure
P	Predicate, positive preference
Q	Quantifier
Q_c	Quality of summary
r	Database tuple
R	Relation, rule, restriction
s	t-conorm or s-norm
s_m	Maximum s-norm
s_a	Algebraic sum
s_L	Łukasiewicz s-norm
s_d	Drastic s-norm
$supp(A)$	Support of fuzzy set
S	Summarizer, simplicity
t	t-norm
t_m	Minimum t-norm
t_p	Product t-norm
t_L	Łukasiewicz t-norm
t_d	Drastic product
t_{nM}	Nilpotent minimum t-norm
T	Transformation
U	Usefulness
v	Validity of linguistic summary
w	Weight
X	Universe of disclosure, universal set
α	And if possible operator
β	Or else operator, threshold
δ	Firing degree of rule
ε	Length of slope of fuzzy set
μ	Membership function
φ	Characteristic function
θ	Length of flat segment of fuzzy set
$\mathbb{N}$	Set of natural numbers
$\mathbb{R}$	Set of real numbers

Chapter 1
Fuzzy Set and Fuzzy Logic Theory in Brief

Abstract A set consists of elements sharing the same property. This property is essential for setting set boundaries. Hence, the following question appears: Can we always unambiguously define these boundaries? The answer is, no. We can unambiguously define a set containing all municipalities belonging to the district D. Municipality either belongs to the district D (from administrative point of view), or does not belong. However, for the set expressing *high distance* we cannot clearly define sharp boundary to distinguish high from non-high distance. This section begins with the classical sets in order to smoothly continue to fuzzy sets. Next, relevant properties and operations of fuzzy sets are discussed. Further, the concept of fuzzy number, as a subcategory of fuzzy sets, is explained. Fuzzy sets and many-valued logics are basis for fuzzy logic. Fuzzy logic facilitates commonsense reasoning with imprecise predicates expressed as fuzzy sets. In the second part fuzzy conjunction, negation, disjunction, implication and quantifiers are examined. Mentioned concepts are used throughout the book.

1.1 From Crispness to Fuzziness

The crisp set is a collection of elements which share the same property. The principal concept in the set theory is belonging or membership to a set. If an element of the universal set X belongs to the set A, we simply write $x \in A$. If x is not a member of A, we write $x \notin A$. It means that belonging to a set should be clear [47].

A crisp set can be described by several methods. The listing method lists all elements by putting them into the braces: $A = \{$strongly agree, agree, do not know, disagree, strongly disagree$\}$, where A denotes all possible answers in a questionnaire, for example. The order of elements is irrelevant. This method is feasible only if a set contains finite number of elements. Otherwise, a set should be described by the membership rule (property or predicate which has to be satisfied)

$$A = \{x \in X \mid x \text{ satisfies property } P\} \tag{1.1}$$

© Springer International Publishing Switzerland 2016
M. Hudec, *Fuzziness in Information Systems*,
DOI 10.1007/978-3-319-42518-4_1

where A denotes a set of all x such that x satisfies property P e.g. $A = \{x \in \mathbb{R} \mid x > 350\}$, where $\mathbb{R}$ is a set of real numbers.

Finally, crisp set A can be defined by the characteristic function φ_A that matches each element of the universal set X to the set A in the following way:

$$\varphi_A(x) : X \rightarrow \{0, 1\} \tag{1.2}$$

Example 1.1 For illustrative example, a governmental agency decided to financially support highly polluted municipalities. In order to discern a set of highly polluted municipalities (HP), the agency should define property explaining the HP set. If high pollution means more or equal than 20 mg of the measured pollutant, then the crisp set HP is defined as: $HP = \{x \in X \mid x \geq 20\}$, where X is the universal set of all possible pollutions (real values greater or equal zero). Membership to a set HP can be expressed as a characteristic function $\varphi(x)$ in the following way:

$$\varphi_{HP}(x) = \begin{cases} 0 & \text{for } x < 20 \\ 1 & \text{for } x \geq 20 \end{cases}.$$

The set expressing high pollution is shown in Fig. 1.1. At the first glance we see the drawback of crisp sets. A municipality having the value of 19.93 mg will not receive any financial support (it is treated in the same way as municipality having value of e.g. 0.2 mg), whereas municipality having pollution of 20 mg will receive full financial support as well as municipality having pollution of 60 mg. Furthermore, measured pollution is expected to be a crisp real number. But in reality, either it is not possible to realize extremely precise measuring [31], or values are fuzzy in their nature. It especially holds for the environmental data [44].   □

Vagueness concerning the description of the semantic meaning of events or phenomena is called fuzziness [57]. Hence, uncertainty is not based on randomness or probability; it cannot be presented as a crisp value. Vague terms such as *high, cheap, medium, around m* ($m \in \mathbb{R}$), *heap* share three interrelated features of vagueness [18]: admit borderline cases, lack sharp boundaries and are susceptible to sorties paradoxes.

Fig. 1.1 Crisp set *high pollution*

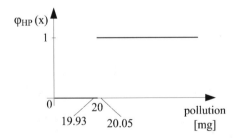

In Fig. 1.1, two kinds of fuzziness are neglected: fuzziness in data and fuzziness in belonging to a set [16]. If managing pollution by sharp sets remains, several small intervals e.g. [0, 15)—no support; [15, 18)—30 % of full support; [18, 22)—65 % of full support; [22, 25)—85 % of full support; etc., to ensure that similar municipalities receive similar support can be employed.

However, when we include additional attributes, such as unemployment or number of respiratory diseases, then managing rules by crisp sets become more complex.

The interpretation of fuzzy sets [56] has arisen from the generalization of the classical sets to embrace the vague notions and unclear boundaries. It may not be always clear, if an element x belongs to a set A, or not. Thus, its membership may be measured by a degree, commonly known as the *membership degree* taking a value from the unit interval by agreement.

Consequently, a fuzzy set A over the universe of discourse X is defined by function μ_A that matches each element of the universe of discourse with its membership degree to the set A

$$\mu_A(x) : X \rightarrow [0, 1] \tag{1.3}$$

where $\mu_A(x) = 0$ says that an element x definitely does not belong to a fuzzy set A, $\mu_A(x) = 1$ says that x without any doubt is member of fuzzy set A. Higher value of $\mu_A(x)$ indicates the higher degree of membership of an element x to a fuzzy set A. Each fuzzy set is defined by one membership function. A membership function maps each element of the universal set X into real numbers from the [0, 1] interval. We should emphasize that the universal set X is always a crisp set [21].

A fuzzy set can be defined as a set of ordered pairs

$$A = \{(x, \mu_A(x)) \mid x \in X \land \mu_A(x) \in (0, 1]\} \tag{1.4}$$

When the universal set is finite, fuzzy set constructed on this universal set can be expressed by counting the elements and their respective membership degrees

$$A = \frac{\mu_A(x_1)}{x_1} + \frac{\mu_A(x_2)}{x_2} + \cdots + \frac{\mu_A(x_n)}{x_n} \tag{1.5}$$

Example 1.2 Let us consider highly polluted municipalities (Example 1.1) from the fuzzy sets point of view. The crisp set HP from Fig. 1.1 could be converted into the fuzzy set FHP in the following way (Fig. 1.2):

$$\mu_{FHP}(x) = \begin{cases} 0 & \text{for } x \leq 15 \\ \dfrac{x - 15}{5} & \text{for } 15 < x < 20 \\ 1 & \text{for } x \geq 20 \end{cases} \tag{1.6}$$

In this way the soft transition between belonging and non-belonging to a set is ensured. □

Fig. 1.2 Fuzzy set *high pollution*

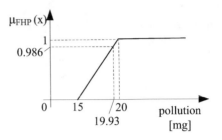

Example 1.3 Let us consider respondent's consent to a question in a questionnaire. The fuzzy set consent (A) can be expressed as A={(strongly agree, 1), (agree, 0.65), (do not know, 0.15)}. Ordered pairs (disagree, 0) and (strongly disagree, 0) are not included into fuzzy set A.   □

1.2   Fuzzy Sets

The first publications of fuzzy set theory by Zadeh [56] and Goguen [11] were focused on the generalization of the classical notion of sets and propositions in order to mathematically cover fuzziness. Although the needs as well as importance of managing fuzziness were recognized earlier (e.g. [39]), the acceptance by scientific and practitioners communities was not high, especially at the beginning. Reasons for accepting and non-accepting fuzzy sets and related topics were summarized in [52].

1.2.1   Properties of Fuzzy Sets

In this section properties relevant for the next sections are examined.

Scalar and relative scalar cardinality
For any fuzzy set A defined on a finite universal set X we define its scalar cardinality by the formula

$$\text{card}(A) = |A| = \sum_{x \in X} \mu_A(x) \tag{1.7}$$

The scalar cardinality of fuzzy set (1.7) is a generalization of the classical cardinality. Elements of universal set belong to the fuzzy sets with different membership degrees and therefore we cannot count elements of a set A, but their respective membership degrees should be summed. Some authors refer to $|A|$ as the *sigma count of* A [21].

The relative scalar cardinality is defined by the formula:

$$||A|| = \frac{\text{card}(A)}{\text{card}(X)} = \frac{|A|}{|X|} \tag{1.8}$$

where $\text{card}(A)$ is defined in (1.7) and $\text{card}(X)$ represents the number of elements in X. These cardinalities are broadly used in e.g. linguistic summaries.

The third type of cardinality is fuzzy cardinality expressed as ordered pair: number of elements belonging to a particular α-cut and α-cut [22] when universal set is a finite one. Cardinalities are closely examined in e.g. [45].

Scalar cardinality of a fuzzy set can be expressed as the area bounded by the membership function of fuzzy set and the x-axis [35]. This approach is demonstrated on the trapezoidal fuzzy set in Sect. 1.2.2.

Support

The support of a fuzzy set A is the crisp set with the following property:

$$\text{supp}(A) = \{(x \in X \mid \mu_A(x) > 0\} \tag{1.9}$$

This property is broadly used in flexible queries, among others.

Core

The core of a fuzzy set A is the crisp set with the following property:

$$\text{core}(A) = \{(x \in X \mid \mu_A(x) = 1\} \tag{1.10}$$

In the fuzzy sets literature the term *kernel* is used as a synonym for the core.

Height

The height is the highest value of membership degree of all elements in the considered fuzzy set A, i.e.

$$h(A) = \sup_{x \in X} \mu_A(x) \tag{1.11}$$

From (1.10) and (1.11) we can infer that if the core is not an empty set, then the height is equal to the value of 1. The opposite does not always hold.

Example 1.4 A heap of maize grains obviously contains large number of grains. Because by crisp sets we cannot unambiguously discern the two sets, *heap* and *non-heap*, a fuzzy set should be applied. For example, one could agree that 2 000 grains is a large quantity (heap), and between 1 and 2 000, the belonging to a heap grows. Thus, the membership function of the set *heap* could be:

$$\mu_{heap}(x) = \begin{cases} 0 & \text{for } x = 0 \\ \frac{2}{\pi} \arctan(0.09 \cdot x) & \text{for } x > 0 \end{cases} \tag{1.12}$$

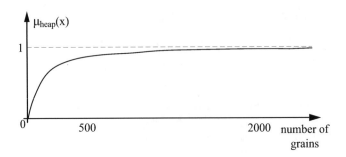

Fig. 1.3 Heap of maize grains explained by fuzzy set

The graphical interpretation of the resulting membership function is shown in Fig. 1.3. This figure explains the core and height properties of fuzzy set. In this case the support (1.9) is unlimited. The height (1.11) is equal to 1, because $\lim_{x \to \infty} \frac{2}{\pi} \arctan(0.09 \cdot x) = 1$. The core (1.10) is an empty set. Although, $h(heap) = 1$, this value is not reached by any element.

Normalized fuzzy set
Fuzzy set A is normalized, if the membership degree of at least one element is equal to 1, i.e.:

$$\exists x \in X, \mu_A(x) = h(x) = 1 \tag{1.13}$$

Crossover point
The element x_{cp} of a fuzzy set A that has a membership degree equal to 0.5 represents the crossover point, i.e.:

$$x_{cp} = \{x \in X | \mu_A(x) = 0.5\} \tag{1.14}$$

α-cut and strong α-cut
One of the important concepts used in fuzzy sets is the α-cut. The α-cut $A^{(\alpha)}$ and its restrictive variant *strong* α-cut $A^{(\alpha+)}$ are defined in the following way:

$$A^{(\alpha)} = \{x \in X \mid \mu_A(x) \geq \alpha\} \tag{1.15}$$

$$A^{(\alpha+)} = \{(x \in X \mid \mu_A(x) > \alpha\} \tag{1.16}$$

where $\alpha \in [0, 1]$.

The α-cut of a fuzzy set A is a crisp set containing all the elements of the X whose membership degrees in A are greater than or equal than the specified value of α. This

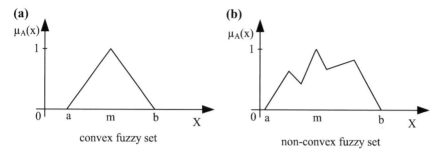

(a)

$\mu_A(x)$

convex fuzzy set

(b)

$\mu_A(x)$

non-convex fuzzy set

Fig. 1.4 Convex and non-convex fuzzy sets

property is used in many directions, e.g. working with elements which significantly belong to a fuzzy set.

Convexity of fuzzy sets

A fuzzy set is convex, if and only if [56]:

$$\mu_A(\lambda x + (1 - \lambda)y) \geq \min(\mu_A(x), \mu_A(y)) \qquad (1.17)$$

for all x and $y \in X$ and all $\lambda \in [0, 1]$. Convex and non-convex fuzzy sets are plotted in Fig. 1.4.

1.2.2   Types of Fuzzy Sets (Membership Functions)

Membership functions are classified into two main groups [10]: linear and Gaussian or curved. All membership functions explained in this section are normalized fuzzy sets.

Triangular fuzzy set (Fig. 1.5) is defined by its lower limit a, its upper limit b and the modal (highest) value m as

Fig. 1.5 Triangular fuzzy set (membership function)

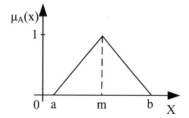

$\mu_A(x)$

Fig. 1.6 Gaussian fuzzy set
(membership function)

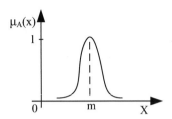

$$\mu_A(x) = \begin{cases} 1 & \text{for } x = m \\ \dfrac{x-a}{m-a} & \text{for } a < x < m \\ \dfrac{b-x}{b-m} & \text{for } m < x < b \\ 0 & \text{for } x \le a \vee x \ge b \end{cases} \qquad (1.18)$$

Gaussian fuzzy set (Fig. 1.6) is defined by the modal value (centre) m and width k as

$$\mu_A(x) = e^{-k(x-m)^2} \qquad (1.19)$$

The bell of the Gaussian function depends on the value k. If the value k is lower, then the bell is narrower.

Trapezoidal fuzzy set (Fig. 1.7) is defined by its lower limit a, its upper limit b, and the flat segment $[m_1, m_2]$ representing the highest value of height (1.11) as

$$\mu_A(x) = \begin{cases} 1 & \text{for } m_1 \le x \le m_2 \\ \dfrac{x-a}{m_1-a} & \text{for } a < x < m_1 \\ \dfrac{b-x}{b-m_2} & \text{for } m_2 < x < b \\ 0 & \text{for } x \le a \vee x \ge b \end{cases} \qquad (1.20)$$

Fig. 1.7 Trapezoidal fuzzy
set

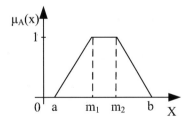

Fig. 1.8 L fuzzy set

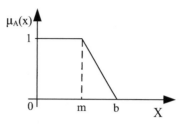

The scalar cardinality (1.7) of trapezoidal fuzzy set is calculated by its area in the following way [35]:

$$\text{card}(A) = \text{area}(A) = (m_2 - m_1) + \frac{(m_1 - a) + (b - m_2)}{2} \tag{1.21}$$

where parameters are the same as in (1.20). In case of triangular fuzzy set the left part is not used.

Trapezoidal, triangular and Gaussian fuzzy sets are suitable for modelling concepts such as *medium value* or *approximate m*, where *m* is a real number. The support (1.9) of the Gaussian fuzzy set is spread over the whole universe of disclosure, although with values close to 0 near the edges of the universe of disclosure (or far from the value of *m*). This could be a problem in fuzzy relational databases, which is discussed later on.

L fuzzy set (Fig. 1.8) is defined by two parameters, *m* and *b*, in the following way:

$$\mu_A(x) = \begin{cases} 1 & \text{for } x \le m \\ \dfrac{b - x}{b - m} & \text{for } m < x < b \\ 0 & \text{for } x \ge b \end{cases} \tag{1.22}$$

L fuzzy set is suitable for defining sets expressing small values of the analysed concepts such as *small pollution*. These concepts can be defined by nonlinear functions as well. Concerning practical applications examined in the next chapters and the simplicity for end users, nonlinear functions are not further considered. Anyway, approaches examined in the book are valid for nonlinear functions. The difference is in calculated values of membership degrees.

R fuzzy set (linear gamma) (Fig. 1.9) is defined by two parameters, *a* and *m*, in the following way:

$$\mu_A(x) = \begin{cases} 0 & \text{for } x \le a \\ \dfrac{x - a}{m - a} & \text{for } a < x < m \\ 1 & \text{for } x \ge m \end{cases} \tag{1.23}$$

Fig. 1.9 R (linear gamma)
fuzzy set

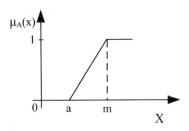

Fig. 1.10 Singleton fuzzy
set

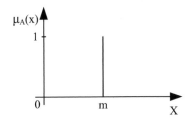

R fuzzy set is suitable for defining sets expressing high values of the analysed concepts, such as *high turnover*. The same comment for nonlinearity of L fuzzy sets holds for the R fuzzy sets.

Singleton fuzzy set (Fig. 1.10) takes the value zero in all the $x \in X$ except in the point $x = m$, where it takes the value 1 in the following way:

$$\mu_A(x) = \begin{cases} 0 & \text{for } x \neq m \\ 1 & \text{for } x = m \end{cases} \tag{1.24}$$

At the first glance, the singleton is an usual crisp number and there is no special need to express crisp number in this way. But, in the tasks of approximate reasoning and in managing fuzziness by relational databases, singletons are indispensable.

According to (1.17) all aforementioned fuzzy sets (1.18)–(1.24) are convex. Furthermore, applying α-cut (1.15) we can say that a fuzzy set A is convex, if and only if all $A^{(\alpha)}$ intervals are convex for $\forall \alpha \in [0, 1]$.

Summarizing this part, we could say that fuzzy sets allow users to express the uncertainty of the analysed problem. On the other hand, the analysed system will not work properly, if membership functions are badly defined. Hence, these functions have to be carefully defined [10].

1.2.3 Operations on Fuzzy Sets

The operations with fuzzy sets A and B are defined via operations on their respective membership functions.

Equality

The fuzzy sets A and B are equal ($A = B$), if for $\forall x \in X$:

$$\mu_A(x) = \mu_B(x) \tag{1.25}$$

This operation is the generalization of equality from classical set theory. However, sets might be more or less equal. For this purpose distance measures and equality indexes (optimistic, medium, pessimistic) are employed [10], as well as generalized equality [34].

Generalized equality generalizes the operator $=$. One of the ways is the generalization of a well known equality relation from the crisp set theory

$$(A \subseteq B \wedge B \subseteq A) \Leftrightarrow A = B \tag{1.26}$$

Straightforwardly, for the fuzzy equality holds: if $(A \subseteq_F B \wedge B \subseteq_F A)$, then $A =_F B$.

Possibility measure

The possibility that the fuzzy value B belongs to a fuzzy concept A is defined as (Fig. 1.11)

$$Poss(B, A) = \sup_{x \in X}[t(\mu_A(x), \mu_B(x))] \tag{1.27}$$

where t stands for t-norm. T-norms are examined in Sect. 1.3.1. Usually minimum t-norm is used and therefore this equation is known as:

$$Poss(B, A) = \sup_{x \in X}[\min(\mu_A(x), \mu_B(x))] \tag{1.28}$$

The possibility measure gets value of 0 when intersection of two fuzzy sets is empty. This consequence is used in fuzzy queries over fuzzy values in the relational databases, among others.

Fig. 1.11 The possibility that the fuzzy value belongs to a fuzzy concept

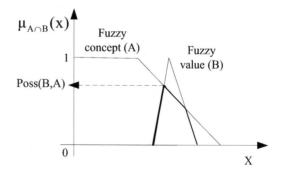

Fig. 1.12 Intersection of
two fuzzy sets

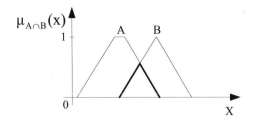

Inclusion

The fuzzy set A is included in the fuzzy set B, if for every $x \in X$ holds

$$\mu_A(x) \le \mu_B(x) \tag{1.29}$$

Hence, fuzzy set A is a subset of fuzzy set B.

Intersection

The intersection operation of fuzzy sets A and B is defined as

$$\mu_{A \cap B}(x) = \min(\mu_A(x), \mu_B(x)) \tag{1.30}$$

Fuzzy sets and their intersection (marked as a bold line) are plotted in Fig. 1.12. If membership degrees are reduced to values of 0 and 1, this function meets the definition of intersection in the classical set theory. Furthermore, this operation is often subnormalized fuzzy set, that is, its height (1.11) is lower than the value of 1, except when $A^{(1)} \cap B^{(1)} \ne \varnothing$. Furthermore, if fuzzy sets A and B are convex, so is their intersection [56].

Union

The union operation of fuzzy sets A and B is defined as

$$\mu_{A \cup B}(x) = \max(\mu_A(x), \mu_B(x)) \tag{1.31}$$

Fuzzy sets and their union (marked as a bold line) are shown in Fig. 1.13. If membership degrees are reduced to values of 0 and 1, this function meets the definition of union in the classical set theory. The union of two fuzzy sets is mainly non-convex, except when the intersection of cores (1.10) is not empty and both sets are convex.

Complement

The fuzzy sets A and $\overline{A}$ are complements if

$$\mu_A(x) = 1 - \mu_{\overline{A}}(x) \tag{1.32}$$

Fig. 1.13 Union of two
fuzzy sets

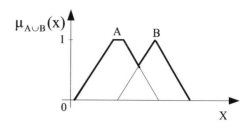

Fig. 1.14 Fuzzy set and its
complement

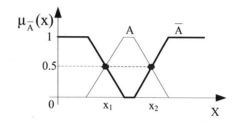

Fuzzy set and its complement are shown in Fig. 1.14. These two sets intersect in the membership degree of 0.5. This point is known as the maximal uncertainty point. For an element having this membership degree we are not sure whether it is more in the set than in its complement. It is obvious, that the axiom of non-contradiction from the crisp set theory (intersection of set and its complement produces an empty set) is not valid in fuzzy sets [36]. Intuitively, violation of this rule seems to be expectable, because uncertainty of belonging to a fuzzy set is reflected in belonging to its complement. On the other hand, there are applications, where violation of this axiom might cause unexpected results.

The fuzzy set theory generalizes the classic set theory. It means that when membership functions (1.3) are reduced to characteristic ones (1.2), all the results are in accordance with the Boolean algebra.

These and other operations are examined in more details in e.g. [7, 10, 21, 32].

1.2.4 Fuzzy Numbers and Fuzzy Arithmetic

A fuzzy number is defined as a convex (1.17) and normalized (1.13) fuzzy set [3]. All fuzzy sets depicted in Figs. 1.5, 1.6, 1.7, 1.8, 1.9 and 1.10 are convex. Obviously, fuzzy numbers are subset of fuzzy sets. In practice, the property of the bounded support (1.9) is relevant. Therefore, triangular and trapezoidal fuzzy sets are fuzzy numbers, whereas Gaussian fuzzy set is not. Membership function should be piecewise continuous and convex. The reason, why fuzzy set should be convex, is easy to prove. Suppose that the fuzzy concept *approximately 100* is expressed as a non-convex fuzzy set shown in Fig. 1.4b ($m = 100$). We expect that membership degrees of elements closer to m are higher than membership degrees of elements farther to m. But, in case of a non-convex fuzzy set it does not hold. Concepts like *approximate 5*, *more or less between 10 and 12* are modelled by fuzzy numbers.

Earlier definition of fuzzy number stated that there should exist exactly one $x_0 \in X$ for which $\mu_A(x_0) = 1$. In this context, value x_0 is called the mean value of A [57]. Nowadays, this requirement is relaxed, allowing trapezoidal functions to express fuzzy numbers.

Comparing to crisp numbers, we can say that a fuzzy number A is positive, if holds $\mu_A(x) = 0$ for $\forall x < 0$. Analogously, a fuzzy number A is negative, if holds $\mu_A(x) = 0$ for $\forall x > 0$.

Example 1.5 People often measure values by estimation. For example, someone could declare that speed was approximately 90 km/h, but for sure not lower than 75 km/h and not higher than 110 km/h. This uncertainty could be managed by triangular fuzzy number in a way that $m = 90$, $a = 75$ and $b = 110$. □

The triangular fuzzy number for the simplicity is denoted by

$$A = (a, m, b) \tag{1.33}$$

in accordance with Fig. 1.5.

In the same way trapezoidal fuzzy number is denoted by

$$A = (a, m_1, m_2, b) \tag{1.34}$$

Operations of addition, multiplication and division are basic ones on crisp numbers, especially when we would like to reveal sums and averages [3]. Therefore, each relational database management system (RDBMS) supports these operations.

The sum of two triangular numbers is also a triangular number

$$\begin{aligned} A_1 + A_2 &= (a^{(1)}, m^{(1)}, b^{(1)}) + (a^{(2)}, m^{(2)}, b^{(2)}) = \\ &= (a^{(1)} + a^{(2)}, m^{(1)} + m^{(2)}, b^{(1)} + b^{(2)}) \end{aligned} \tag{1.35}$$

The sum of two trapezoidal numbers is also a trapezoidal number

$$\begin{aligned} A_1 + A_2 &= (a^{(1)}, m_1^{(1)}, m_2^{(1)}, b^{(1)}) + (a^{(2)}, m_1^{(2)}, m_2^{(2)}, b^{(2)}) = \\ &= (a^{(1)} + a^{(2)}, m_1^{(1)} + m_1^{(2)}, m_2^{(1)} + m_2^{(2)}, b^{(1)} + b^{(2)}) \end{aligned} \tag{1.36}$$

Triangular number can be expressed as $A = (a, m, m, b)$ which allows us to realize addition of triangular and trapezoidal fuzzy numbers.

Multiplication of a triangular number by a real number p is also a triangular number

$$p \cdot A = p(a, m, b) = (p \cdot a, p \cdot m, p \cdot b) \tag{1.37}$$

Division of a triangular number A by a real number p is defined as multiplication of A by $\frac{1}{p}$, $p \neq 0$

$$\frac{A}{p} = \frac{1}{p} \cdot A = \frac{1}{p}(a, m, b) = (\frac{a}{p}, \frac{m}{p}, \frac{b}{p}) \qquad (1.38)$$

Analogously, division of a trapezoidal number by a real number p is defined as

$$\frac{A}{p} = \frac{1}{p}(a, m_1, m_2, b) = (\frac{a}{p}, \frac{m_1}{p}, \frac{m_2}{p}, \frac{b}{p}) \qquad (1.39)$$

Now, we are able to calculate averages (and another operations) of fuzzy numbers expressed as triangular and trapezoidal sets.

If the supports of L and R fuzzy sets are bounded, then they are fuzzy numbers on which we can apply arithmetic operations. L fuzzy set can be expressed as $A = (m_1, m_1, m_2, b)$. Otherwise, the support of resulting set is not bounded.

Example 1.6 Let us have a triangular fuzzy set $A_1(8, 10, 12)$ and a R fuzzy set $A_2(20, 25, \infty, \infty)$. The sum of these two fuzzy sets is obtained by (1.36), when A_1 is converted into trapezoidal fuzzy set, as $A(28, 35, \infty, \infty)$, when bounded support is excluded from the properties of fuzzy numbers. $\qquad \square$

1.2.5 Measures of Fuzziness

This measure indicates the degree of fuzziness of fuzzy sets, i.e. how far is a fuzzy set from a crisp one. Generally, entropy and degree of distinction between the fuzzy set and its complement are used to calculate measure of fuzziness.

In Sect. 1.1 general explanation of fuzziness is provided. Further explanations of fuzziness uses particular parameters of sets such as: lack of distinction between a fuzzy set and its complement in sense of Zadeh [49] or entropy represented by the uncertainty related to the corresponding crisp set [46].

Let $\mu_A(x)$ be the membership function of fuzzy set A defined on the finite universe of disclosure X. The measure of fuzziness $d(A)$ has the following properties [25]:

- P1: $d(A) = 0$, if A is a crisp set (subset in X)
- P2: $d(A)$ assumes a unique maximum, if $\mu_A(x) = 0.5$ for $\forall x \in X$
- P3: $d(A) \geq d(B)$, that is, B is crisper than A, if $\mu_B(x) \leq \mu_A(x)$ for $\mu_A(x) \leq 0.5$ and $\mu_B(x) \geq \mu_A(x)$ for $\mu_A(x) \geq 0.5$
- P4: $d(\overline{A}) = d(A)$.

Property P3 is also expressed as $A \leq_s B \Leftrightarrow \min(0.5, \mu_A(x)) \geq \min(0.5, \mu_B(x)) \wedge \max(0.5, \mu_A(x)) \leq \max(0.5, \mu_B(x))$, where $\leq_s$ is relation *less crisp than*. This property is illustrated in Fig. 1.15.

Fig. 1.15 Measure of
fuzziness by property P3:
fuzzy set A is less crisp than
fuzzy set B

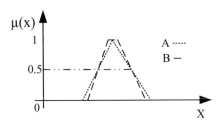

The measure which meets these properties of fuzziness is based on the entropy
and defined as [25]

$$d(A) = H(A) + H(\overline{A}), x \in X$$
$$H(A) = -k \sum_{i=1}^{n} \mu_A(x_i) \ln \mu_A(x_i) \tag{1.40}$$

where n is the number of elements in the support of A and $k > 0$. Analogously,
$H(\overline{A}) = -k \sum_{i=1}^{n} \mu_{\overline{A}}(x_i) \ln \mu_{\overline{A}}(x_i)$. Applying Shannon's function of entropy
$S(x) = -x \ln x - (1 - x) \ln(1 - x)$ yields

$$d(A) = k \sum_{i=1}^{n} S(\mu_A(x_i)) \tag{1.41}$$

Example 1.7 Let A be fuzzy set expressing *integers close to 6* as
$A = \{(4, 0.4), (5, 0.85), (6, 1), (7, 0.85), (8, 0.4)\}$
and B be fuzzy set expressing *integers fairly close to 6* as
$B = \{(3, 0.15), (4, 0.35), (5, 0.75), (6, 1), (7, 0.75), (8, 0.35), (9, 0.15)\}$.

For $k = 1$ we compute the following measures of fuzziness: $d(A) = 2.2244$
and $d(B) = 3.265$. The result clearly shows that more vague fuzzy set has higher
measure of fuzziness. Thus, fuzzy set B is more vaguely defined than fuzzy set A.
If all membership degrees in B are equal to 0.5, the result is $d(B) = 4.8524$. □

The second measure is focused on the lack of distinction between fuzzy set and
its complement. In the case of crisp sets $A \cap \overline{A} = \emptyset$ always holds. It means that these
sets are clearly distinct. If $\mu_A(x) = \frac{1}{2}$ for $\forall x \in X$, then according to the property P2
fuzzy set and its complement are equal.

Mathematically, measure of fuzziness can be defined as [49]:

$$f(A) = 1 - \frac{D_p(A, \overline{A})}{||\text{supp}(A)||} \tag{1.42}$$

where $||\text{supp}(A)||$ is a relative cardinality (1.8) of the support of fuzzy set A and
D_p is a distance between set and its complement, i.e. $D_p(A, \overline{A}) = \left[\sum_{i=1}^{n} |\mu_A(x_i) - \mu_{\overline{A}}(x_i)|^p \right]^{1/p}$, where p is a natural number. For $p = 1$ we get the Hamming metric,
for $p = 2$ we get the Euclidean metric and so forth.

Properties P2 and P3 are not suitable for measuring fuzziness, where membership degree of each element contributes individually, i.e. it is not related to other elements. An example is a fuzzy set expressing the foreign language competency of students. For example, the student S speaks 1/French, 0.75/German, 0.25/Spanish. We think that this student speaks better French than German. Students can speak all languages excellently or on a medium level (0.5). In order to solve this issue new P2 and P3 measures are suggested in [40].

Measures of fuzziness are further discussed in Sects. 1.4 and 1.5. In the former, this measure may express quality of linguistic terms in IF-THEN rules. In the latter, this information is useful to know for decision, whether fuzzy data could be stored in a database, or validation is required, for example.

1.2.6 Fuzzy Relations

A fuzzy relation on the Cartesian product $A \times B$ is defined as set:

$$R = \{((x, y), \mu_R(x, y)) \mid \forall (x, y) \in A \times B, \mu_R(x, y) > 0\} \qquad (1.43)$$

Fuzzy relation has stronger expressive power than the crisp counterpart. By fuzzy relation it is possible to relate elements of two sets by linguistic relations such as *x more or less similar than y* or *x significantly greater than doubled y*. The former fuzzy relation can be expressed by membership function

$$\mu_R(x, y) = \begin{cases} 0 & \text{for } x \leq 0.7y \vee x \geq 1.3y \\ 1 - |x - y| & \text{for } x > 0.7y \wedge x < 1.3y \\ 1 & \text{for } x \geq 0.95y \vee x \leq 1.05y \end{cases} \qquad (1.44)$$

Example 1.8 Let us consider $A = \{x_1, x_2, x_3, x_4\}$ and $B = \{y_1, y_2, y_3\}$ to be sets of towns. Set A represents towns, where storages are located and B towns, where shops are located. Short distance between these towns is expressed as relation $\mu_R(x, y)$ shown in Table 1.1.   □

Table 1.1 Short distances between towns

	y_1	y_2	y_3
x_1	0	0.20	0.90
x_2	0.35	0.45	0.15
x_3	0.20	1	1
x_4	0.55	0.35	1

Basic operations on fuzzy relations are the same as for fuzzy sets, because the fuzzy relation is expressed as a set. More about these operations (equality, inclusion, intersection, etc.) is in e.g. [3, 21, 57].

Compositions of fuzzy relations

Fuzzy relations in different Cartesian product spaces can be merged by composition operations. The best known is the max-min composition [57]. The max-min composition of relations $R(A_1, A_2)$ and $S(A_2, A_3)$ is expressed by formula

$$\mu_{R\circ S}(x_1, x_3) = \sup_{x_2 \in A_2} \min[\mu_R(x_1, x_2), \mu_S(x_2, x_3)], \quad \forall(x_1, x_3) \in A_1 \times A_3 \quad (1.45)$$

Max-min compositions have their own properties which are mentioned here. More details can be found in e.g. [37, 50].

Associativity

The max-min composition of relations R, S and T is associative: $(R \circ S) \circ T = R \circ (S \circ T)$.

Reflexivity

This property can be divided into two subproperties: relation R is reflexive in the classical sense, when $\mu_R(x, x) = 1$ for $\forall x \in X$; relation R is ε-reflexive, when $\mu_R(x, x) \geq \varepsilon$ for $\forall x \in X$. The parameter ε fuzzifies the reflexivity. Furthermore, if relations R and S are reflexive, then the max-min composition $R \circ S$ is also reflexive.

Symmetry

A fuzzy relation R is symmetric, if $\mu_R(x, y) = \mu_R(y, x)$ for $\forall x, y \in X$.
A fuzzy relation is asymmetric, if for $x \neq y$ either $\mu_R(x, y) \neq \mu_R(y, x)$ or $\mu_R(x, y) = \mu_R(y, x) = 0$ holds.

Transitivity

A fuzzy relation R is (max-min) transitive, if $R \circ R \subseteq R$.

The more general definition of composition is: max-* for the relations on finite sets and sup-* for the relations on infinite sets. The asterix sign stands for different operators, such as minimum or product t-norms [12]. These compositions are further examined in the chapter devoted to the fuzzy reasoning.

1.3 Fuzzy Logic

Fuzzy logic is an extension of the many-valued logic by incorporating fuzzy sets into the system of many-valued logic [3].

In Sect. 1.2.3 we have discussed operations of intersection (1.30), union (1.31) and complement (1.32). Fuzzy sets in these operations should be defined on the same universal set. The result is projected to the same universal set. In the logic we are not limited to a single universal set.

Operations and properties on sets and logic are isomorph, i.e. belonging to a set is equivalent with the statement's truth value, union is equivalent with the disjunction, intersection is equivalent with the conjunction and complement is equivalent with the negation in the following way:

$$x \in A \bigcap B \Leftrightarrow (x \in A \land x \in B)$$
$$x \in A \bigcup B \Leftrightarrow (x \in A \lor x \in B) \qquad (1.46)$$
$$x \in \overline{A} \Leftrightarrow \neg(x \in A)$$

where A and B can be defined on different universal sets.

Accordingly, in the fuzzy logic three main operations are conjunction, disjunction and negation. First, we have to define proposition and predicate which are relevant for further reading.

Proposition is a declarative sentence that is either true (denoted by 1), or false (denoted by 0). Philosophers and mathematicians have always had doubt, how to describe phenomena of real world with only two truth values [2, 6, 39]. When we shift from two-valued to many-valued logics, the proposition could become partially true, that is, truth value is not limited to the values of 1 and 0. First, three-valued logic has been examined. Continuously, many-valued logics have been derived. When truth values were expressed by real numbers from the [0, 1] interval, the infinite valued logics appeared [2]. One of the many-valued logics or infinite valued logics is fuzzy logic.

Proposition is either simple (elementary or atomic), or compound (consists of two or more atomic propositions joined by logical connectives).

Predicate is a declarative sentence containing one or more variables and unknowns. A predicate $P(x)$ is not a proposition, because x is unknown. The predicate becomes a proposition, when x gets a particular value. If a proposition is constructed in the frame of two-valued logic, then the truth value can be either 0 or 1. Opposite, in the fuzzy logic frame the truth value can be any real number from the unit interval.

1.3.1 Fuzzy Conjunction

A suitable tool for the interpretation of the *and* connective (conjunction) in fuzzy logic are triangular norms (or short t-norms) [15]. The concept of triangular norms is based on the idea of probabilistic metric spaces [41] introduced in [26]. T-norm is a binary operation t on the interval [0, 1] ($t : [0, 1]^2 \rightarrow [0, 1]$) which is commutative, associative, monotone and meets boundary axiom (1 as a neutral element) [7, 19, 48]. Each function which meets these four properties is a t-norm. Relevant mathematical aspects of t-norms are deeply discussed in [20]. This section is rather focused on aspects which are relevant for applicability in next sections.

Theoretically, unlimited number of t-norms exists. The four basic and remarkable t-norms are [19]

- minimum

$$t_m(\mu_{A_1}(x), \mu_{A_2}(x)) = \min(\mu_{A_1}(x), \mu_{A_2}(x)) \tag{1.47}$$

- product

$$t_p(\mu_{A_1}(x), \mu_{A_2}(x)) = \mu_{A_1}(x) \cdot \mu_{A_2}(x) \tag{1.48}$$

- Łukasiewicz t-norm

$$t_L(\mu_{A_1}(x), \mu_{A_2}(x)) = \max(0, \mu_{A_1}(x) + \mu_{A_2}(x) - 1) \tag{1.49}$$

- drastic product

$$t_d(\mu_{A_1}(x), \mu_{A_2}(x)) = \begin{cases} 0 & \text{for } (\mu_{A_1}(x), \mu_{A_2}(x)) \in [0, 1)^2 \\ \min(\mu_{A_1}(x), \mu_{A_2}(x)) & \text{otherwise} \end{cases} \tag{1.50}$$

where $\mu_{Ai}(x)$, $i = 1, 2$ denotes the membership degree of the element x to the fuzzy sets A_i.

The associative axiom ensures that all t-norm functions can be extended to n propositions by induction [19]:

$$t_{i=1}^n(\mu_{A_i}(x)) = \begin{cases} 1 & \text{for } n = 0 \\ t(\mu_{A_n}(x), t_{i=1}^{n-1}(\mu_{A_i}(x))) & \text{for } n > 0 \end{cases} \tag{1.51}$$

which is a very desirable property for variety of tasks. From the practical point of view, it is not easy to use all t-norm functions, when we have higher number of atomic propositions. Aforementioned four remarkable t-norms are easy to use in this case.

The set of all possible t-norms is bounded by the largest minimum t-norm and the smallest drastic product, i.e. $t_d \le t \le t_m$, where t is an arbitrary t-norm. Concerning the aforementioned basic t-norms, we can create the following relation:

$$t_d \le t_L \le t_p \le t_m \tag{1.52}$$

where the equation holds when truth values are limited to the set $\{0, 1\}$.

Furthermore, from the basic t-norms only the drastic product (1.50) is a noncontinuous function. For a number of reasons, continuous t-norms play an important role in theory and applications. But, t_d is also applicable.

In the frame of algebra, t is a t-norm, if and only if ([0, 1], t, $\le$) is a totally ordered commutative semigroup with neutral element 1 and annihilator 0 [19]. Hence, algebraic properties of t-norms may beuseful in many tasks related to mining knowledge from the data. In this section we focus on presence of idempotent and nilpotent elements and limit property.

The idempotency property says:

$$\forall \mu_A(x) \in [0, 1] \text{ holds } t(\mu_A(x), \mu_A(x)) = \mu_A(x) \tag{1.53}$$

The property of nilpotency says:

An element $\mu_A(x) \in (0, 1)$ is a nilpotent element of t-norm t, if there exist some $n \in \mathbb{N}$ such that $t^{(n)}(\mu_A(x)) = 0$ for all $\mu_A(x) \in (0, 1)$.

The t-norm has a limit property if for: $\forall x \in (0, 1)$ $\lim_{n \to \infty} t^{(n)}(\mu_A(x)) = 0$.

It is obvious that only the minimum t-norm (1.47) meets the idempotency property. Other t-norms are idempotent only for membership degree equal to 0 and 1. Łukasiewicz t-norm (1.49) and drastic product (1.50) contains nilpotent elements whereas product t-norm meets the limit property.

An interesting t-norm is the nilpotent minimum t-norm defined as [33]:

$$t_{Nm}(\mu_{A_1}(x), \mu_{A_2}(x)) = \begin{cases} 0 & \text{for } \mu_{A_1}(x) + \mu_{A_2}(x) \leq 1 \\ min(\mu_{A_1}(x), \mu_{A_2}(x)) & \text{otherwise} \end{cases}$$

$$\tag{1.54}$$

The set of $\{0\} \cup (0.5, 1]$ is a set of idempotent elements and the set of $(0, 0.5]$ is a set of nilpotent elements. Hence, this t-norm can be considered in applications as idempotent conjunction, when the sum of truth values of both atomic predicates is significant. This property provides benefit which is not possible to obtain with threshold value applied on minimum t-norm. Only membership degrees the sum of which is greater than 1, are considered, similarly as for Łukasiewicz t-norm, but with membership degree equal to minimum t-norm. To illustrate this consideration, differences between minimum t-norm, Łukasiewicz t-norm and nilpotent minimum t-norm are plotted in Fig. 1.16.

minimum t-norm   Łukasiewicz t-norm   nilpotent minimum t-norm

Fig. 1.16 3D graph of minimum (1.47), Łukasiewicz (1.49) and nilpotent minimum (1.54) t-norms

1.3.2 Fuzzy Negation

Each function which meets two basic axioms: boundary condition and monotonicity, is considered as fuzzy negation [21]. Furthermore, a strictly decreasing function is called a strict negation. A strict negation is strong, if it meets the involution axiom, i.e. $c(c(\mu_A(x))) = \mu_A(x)$ for all $\mu_A(x) \in [0, 1]$.

Another relevant property of the strict negation is, that there exists a unique value $\mu_A(x) \in (0, 1)$, for which holds $c(\mu_A(x)) = \mu_A(x)$ [8]. This corresponds with the maximal uncertainty point ($x = 0.5$) of the fuzzy set and its complement plotted in Fig. 1.14.

The standard negation, which is often used, is defined as:

$$c_s(x) = 1 - \mu_A(x) \tag{1.55}$$

Applying standard negation and continuous t-norms without nilpotent element 0.5 we conclude that the law of non-contradiction from the two-valued logic ($p \wedge \neg p$) is violated. If $p = 0.25$, then $1 - p = 0.75$ and therefore $t_m(p, 1 - p) = 0.25$ and $t_p(p, 1 - p) = 0.1875$. On the other hand, this property is satisfied by the Łukasiewicz t-norm.

Similarly as for t-norms, there exist two functions which represent limited values for all negations. They are Gödel negations defined as:

$$c_g(x) = \begin{cases} 1 & \text{for } \mu_A(x) = 0 \\ 0 & \mu_A(x) \in (0, 1] \end{cases} \tag{1.56}$$

and its dual negation

$$c_{dg}(x) = \begin{cases} 1 & \text{for } \mu_A(x) \in [0, 1) \\ 0 & \mu_A(x) = 1 \end{cases} \tag{1.57}$$

Hence, the set of all possible negations is bounded in the following way:

$$c_g \leq c \leq c_{dg} \tag{1.58}$$

where c is an arbitrary negation.

1.3.3 Fuzzy Disjunction

The s-norm or t-conorm functions define a general class of disjunction operators. T-conorm is a binary operation s on the interval $[0, 1]$ ($s : [0, 1]^2 \rightarrow [0, 1]$) which is commutative, associative, monotone and meets boundary axiom (0 as a neutral element) [7, 19]. Each function which meets these four axioms is a s-norm.

S-norm functions can be created axiomatically or as dual functions of t-norms using the well known De Morgan statement: $s(x, y) = 1 - t(1 - x, 1 - y)$ and suitable negation operator, such as the standard negation (1.55) [4]. The name "t-conorm" expresses the fact that these functions are dual functions to t-norms realized by complement. The triple (t, s, c) is called a De Morgan triple [47], where t, s and c stand for t-norm, s-norm and strict negation consequently. This statement helps us to recognize axiomatic and algebraic properties of dual function, i.e. if t is left-continuous then s preserves this property.

The following s-norm functions are correspondingly dual to the aforementioned t-norms (1.47)–(1.50):

- maximum

$$s_m(\mu_{A_1}(x), \mu_{A_2}(x)) = \max(\mu_{A_1}(x), \mu_{A_2}(x)) \qquad (1.59)$$

- algebraic sum

$$s_a(\mu_{A_1}(x), \mu_{A_2}(x)) = \mu_{A_1}(x) + \mu_{A_2}(x) - \mu_{A_1}(x) \cdot \mu_{A_2}(x) \qquad (1.60)$$

- Łukasiewicz s-norm

$$s_L(\mu_{A_1}(x), \mu_{A_2}(x)) = \min(1, \mu_{A_1}(x) + \mu_{A_2}(x)) \qquad (1.61)$$

- drastic s-norm

$$s_d(\mu_{A_1}(x), \mu_{A_2}(x)) = \begin{cases} 1 & \text{for } (\mu_{A_1}(x), \mu_{A_2}(x)) \in [0, 1)^2 \\ \max(\mu_{A_1}(x), \mu_{A_2}(x)) & \text{otherwise} \end{cases}$$

$$(1.62)$$

where $\mu_{A_i}(x)$, $i = 1, 2$ denotes the membership degree of the element x to the fuzzy sets A_i.

Concerning s-norms, only the maximum s-norm is idempotent (it is dual function of the minimum t-norm, which is the only idempotent t-norm).

1.3.4 Fuzzy Implication

Atomic propositions in conjunctions and disjunctions can be independent, e.g. *altitude is small and ratio of public greenery is high*. We just simply search for municipalities which have small altitude and high ratio of parks and other green areas. In the implication atomic propositions should be in causal relationship, e.g. *short distance causes strong pressure on brake pedal*. Even if data mining techniques reveal relationship *if altitude is small then ratio of public greenery is high*, care should be taken before declaring that this is a causal relationship relevant for the policy making.

In the theory of fuzzy logic four models for implication operators exist [43]: strong or S-implications, quantum implications or Q-implications, residuated or R-implications and Mamdani–Larsen or ML-implications.

S-implications are generalized from the tautology $x \Rightarrow y \Leftrightarrow \neg x \vee y$ [42, 51]. In the fuzzy logic disjunction and negation can be expressed with variety of functions. Therefore, we cannot say that only one S-implication exists. For the maximum s-norm (1.59) and standard negation (1.55) we obtain the Kleene-Dienes implication:

$$i_{sKD}(\mu_{A_1}(x), \mu_{A_2}(x)) = \max((1 - \mu_{A_1}(x)), \mu_{A_2}(x)) \qquad (1.63)$$

For the Łukasiewicz s-norm (1.61) and standard negation (1.55) we get the Łukasiewicz implication

$$i_{sL}(\mu_{A_1}(x), \mu_{A_2}(x)) = \min(1, 1 - \mu_{A_1}(x) + \mu_{A_2}(x)) \qquad (1.64)$$

Similarly, Q-implications are generalized from the tautology: $x \Rightarrow y \Leftrightarrow \neg x \vee (x \wedge y)$ [42, 51]. For the maximum s-norm (1.59), minimum t-norm (1.47) and standard negation (1.55) we obtain the Zadeh implication:

$$i_{qZ}(\mu_{A_1}(x), \mu_{A_2}(x)) = \max(1 - \mu_{A_1}(x), \min(\mu_{A_1}(x), \mu_{A_2}(x))) \qquad (1.65)$$

Residuated or R-implications derive from t-norms by residuation in the following way [1, 13]

$$i_r(x, y) = \sup_{c}\{c \in [0, 1], t(a, c) \le b\}, \quad \forall a, b \in [0, 1] \qquad (1.66)$$

For the minimum t-norm (1.47) we obtain the Gödel implication

$$i_{rGd}(\mu_{A_1}(x), \mu_{A_2}(x)) = \begin{cases} 1 & \text{for } \mu_{A_1}(x) \le \mu_{A_2}(x) \\ \mu_{A_2}(x) & \text{for } \mu_{A_1}(x) > \mu_{A_2}(x) \end{cases} \qquad (1.67)$$

These equations say that when $a < b$, the maximal value of c for which $t(a, c) \le b$ holds is 1. When $a > b$ the maximal value of c is equal to b.

For the product t-norm (1.48) we obtain the Goguen implication

$$i_{rGg}(\mu_{A_1}(x), \mu_{A_2}(x)) = \begin{cases} 1 & \text{for } \mu_{A_1}(x) \le \mu_{A_2}(x) \\ \frac{\mu_{A_2}(x)}{\mu_{A_1}(x)} & \text{for } \mu_{A_1}(x) > \mu_{A_2}(x) \end{cases} \qquad (1.68)$$

Apparently, there are many other feasible fuzzy implications discussed in e.g. [9, 29]. Apart from these implications, some applications employ for the implication purposes functions which do not meet all the axioms to be real implications. In practice, it is very usual to describe implication by t-norms [14]. This particularly holds for the minimum t-norm (1.47) which is often called Mamdani implication. The reason is explained in the chapter dedicated to fuzzy reasoning (Sect. 4.2.1).

1.4 Linguistic Variables

Variables whose values are words of natural language, mainly combination of adverbs (e.g. very, few) and adjectives (e.g. small, medium, high) are called linguistic variables. Every value of a linguistic variable represents a fuzzy set [27]. In the terminology of relational databases, variables are entities' attributes such as age, length of roads, pollution. On the domain (or universe of disclosure) of attribute, fuzzy sets for each term are constructed.

Mathematically, a linguistic variable takes values (linguistic labels or terms) expressed in natural language instead of numbers [55]. A linguistic variable is determined by a quintuple $(L, T(L), X, G, H)$, where

- L is the name of the variable
- $T(L)$ is a set of all linguistic labels related to variable L
- X is the universe of discourse of the variable
- G is the syntactic rule to generate $T(L)$ values
- H is the semantic rule that relates each linguistic label of $T(L)$ to its meaning $H(L)$, where $H(L)$ is a diffuse subset of X

Example 1.9 To illustrate the concept of linguistic variable consider the variable *number of days with snow coverage* for a particular year or as average for a longer period. The domain or the universe of discourse (set of integers from the [0, 365] interval) is divided into five fuzzy sets (linguistic labels) shown in Fig. 1.17. Finally each term is linked to its meaning.

The membership function for the term *medium* is

$$\mu_{medium}(x) = \begin{cases} 1 & \text{for } 156 \leq x \leq 208 \\ \dfrac{x - 130}{26} & \text{for } 130 < x < 156 \\ \dfrac{234 - x}{26} & \text{for } 208 < x < 234 \\ 0 & \text{for } x \leq 130 \vee x \geq 234 \end{cases} \quad (1.69)$$

Consequently, other terms are constructed. Furthermore, this variable has different meaning for people looking for an interesting skiing destination and for local authorities planning winter road maintenance needs, for example. □

Flats in Fig. 1.17 express areas, where belonging to a set is unambiguous. Slopes depict ambiguity of being in a set. The maximal value of the intersection between two sets represents the maximal uncertainty degree. The value of 143 has the same membership degree to fuzzy sets *small* and *medium*. If we examine linguistic variable

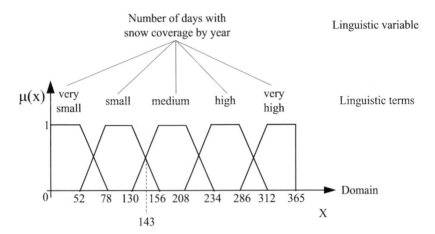

Fig. 1.17 Terms of the linguistic variable *number of days with snow coverage* over the domain [0, 365] of integers

such as distance or turnover, then the term *very high* can be expressed as R fuzzy set (Fig. 1.9).

Linguistic variables play an important role in many fields such as data summarization and inference rules. Not only variables of real world such as distance and temperature, but also quantifiers can be expressed as linguistic variables.

Partition of a universal set X is a grouping of the elements x into subsets, in such a way that every element x is included in only one of the subsets. Union of subsets is the whole set X $(A_1 \bigcup A_2 \bigcup ... \bigcup A_n = X)$ and intersection of each two subsets produces empty set $(A_i \bigcap A_j = \emptyset, \forall i, j \in \{1, 2, ..., n\}, i \neq j)$.

In fuzzy set theory element can belong to several sets with different or equal membership degrees. According to [38] a k-tuple of fuzzy sets $(A_1,..., A_k)$ is a fuzzy partition of X if $\emptyset \neq A_i \neq X, \forall i \in \mathbb{N}_k$ and

$$\sum_{i=1}^{k} \mu_{A_i}(x) = 1, \ \forall x \in X \tag{1.70}$$

Further requirement is the normality of fuzzy sets [30], i.e. height (1.11) of each A_i is equal to 1.

By merging the concept of linguistic variable with fuzzy partitions we can create a family of fuzzy sets of good quality.

Example 1.10 Theoretically, linguistic variable can be of structure shown in Fig. 1.18. But this variable is not a fuzzy partition of the universal set X as is linguistic variable shown in Fig. 1.17. Clearly, for $x = 25$ we get $\sum_{i=1}^{k} \mu_{A_i}(x) = 1.55 > 1$ and for $x = 57$, $\sum_{i=1}^{k} \mu_{A_i}(x) = 0.93 < 1$.

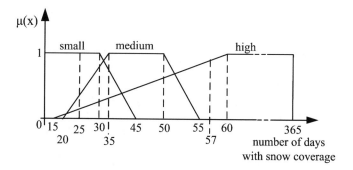

Fig. 1.18 Linguistic variable *number of days with snow coverage* which is not a fuzzy partition of the domain [0, 365]

This linguistic variable is not a fuzzy partition, and thus not very suitable to be used in mining summaries from data or in fuzzy rule base, whereas linguistic variable plotted in Fig. 1.17 is.   □

Concerning the measure of fuzziness, fuzzy sets plotted in Fig. 1.17 have equal degree of fuzziness, whereas fuzzy sets shown in Fig. 1.18 have different degrees of

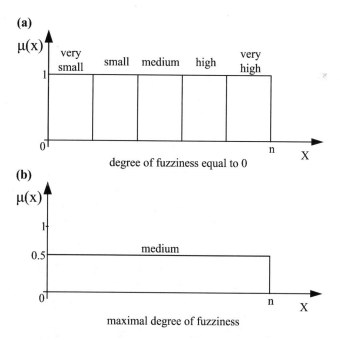

Fig. 1.19 Linguistic variable expressed by partition of crisp sets (degrees of fuzziness = 0 for all sets) and by maximal degree of fuzziness

fuzziness. By this measure we can see whether the family of fuzzy sets is uniformly distributed on set X.

Two extreme situations are plotted in Fig. 1.19. In the upper part the universe of disclosure is partitioned into five crisp sets causing that sum of degrees of fuzziness for all sets is equal to 0 (1.41, 1.42). In the lower part one fuzzy set with membership degree of 1/2 for all $x \in X$ spreads over the X expressing maximal semantic uncertainty (degree of fuzziness). It is more obvious when we construct a complement of this fuzzy set because distance between fuzzy set and its complement is equal to 0 and therefore fuzziness is maximal.

1.5 Fuzzy Quantifiers

Fuzzy or linguistic quantifiers [23, 24, 54] allow us to reveal an approximate idea of the number of elements of a fuzzy set fulfilling a certain proportion in relation to the total number of elements in a universal set. Fuzzy quantifiers can be absolute or relative [10].

Absolute quantifiers express the amount of elements from a particular set which meet the propositions such as *much more than 30 elements, approximately 100 elements* and the like. The function of quantifier is

$$Q_{abs}(x) : \mathbb{R} \rightarrow [0, 1] \qquad (1.71)$$

The truth value of the absolute quantifier gets values from the unit interval.

Relative quantifiers express the proportion of elements from a particular set which meet the propositions such as *most of customers meet P*. Relative fuzzy quantifiers are also expressed as fuzzy numbers. The function of quantifier is:

$$Q_{rel}(x) : [0, 1] \rightarrow [0, 1] \qquad (1.72)$$

where the domain of Q_{rel} is [0,1] because the division of elements which meet the fuzzy proposition and total number of elements gets value from the [0, 1] interval. Hence, relative fuzzy quantifiers are expressed as fuzzy numbers.

The possible membership functions of quantifiers *few, about half, most of* and *almost all* are depicted in Fig. 1.20. The universe of discourse X of these quantifiers is the [0, 1] interval. Parameters l, o, p, s, m and n could be adjusted independently for each quantifier and task. Furthermore, quantifier *about half* can be expressed as triangular or trapezoidal fuzzy set. On the other hand, relative quantifiers can be constructed with the help of linguistic variables and fuzzy partitioning examined in Sect. 1.4 as a family of linguistic terms uniformly distributed on the [0, 1] interval [17]. The family of three quantifiers is plotted in Fig. 1.21.

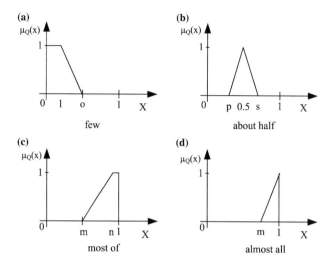

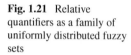

Fig. 1.20 Fuzzy sets describing relative quantifiers. a few; b about half; c most of; d almost all

Fig. 1.21 Relative
quantifiers as a family of
uniformly distributed fuzzy
sets

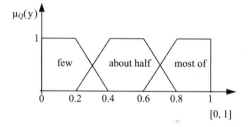

1.6 Some Remarks

The following thoughts by Professor Zadeh explain usefulness of this approach [53]:

> Soft computing is capable to exploit the tolerance for uncertainty and partial truth to achieve robustness, low solution-cost and better rapport with reality solutions in comparison with traditional hard computing.

Even though the fuzziness is closely related to phenomena in social sciences and business, the mathematics of fuzzy logic is mainly applied in engineering and computer science [28]. This trend continues allowing technical systems to be more and more sophisticated and powerful. We could reach the same in social sciences and business, if we efficiently support them by fuzzy logic.

The fact that a predicate and its negation may be true to some extent, is an essence of fuzzy logic. We lose some properties (excluded middle, non-contradiction, idempotency), when we generalize two-valued logic to fuzzy logic. On the other hand, we get a more powerful tool for analysing real world tasks [15]. However, we have to be careful when these properties should be satisfied. Issues which may appear in fields which are covered by this book are discussed and solutions are provided.

The interpolative realization of Boolean algebra (IBA) [36] brings fuzziness into the frame of the Boolean algebra. The IBA consists of symbolic or qualitative level (related to finite Boolean algebra), where all properties of Boolean algebra are satisfied, and of semantic or valued level (a matter of interpretation where truth values from the [0, 1] interval are applied). Moreover, IBA is in the line with the ideas expressed in [5], where it is stated that "... the symbols of the (logic) calculus do not depend for their interpretation upon the idea of quantity ..." and only "in their particular application ..., conduct us to the quantitative conditions of inference".

The following statement nicely explains fuzzy logic and topics discussed in this book

> Fuzzy logic focuses on linguistic variables in natural language and provides a foundation for approximate reasoning with imprecise propositions. It reflects both the rightness and vagueness of human thinking [6].

In the next sections examples are mainly demonstrated on the illustrative data. More complex examples which include building and adjusting interfaces are realized on the statistics database of territorial units. These examples are in appendixes.

References

1. Baczyński, M., Jayaram, B.: (S, N)-and R-implications: A state-of-the-art survey. Fuzzy Sets Syst. **159**, 1836–1859 (2008)
2. Bergmann, M.: An Introduction to Many-Valued and Fuzzy Logic—Semantics, Algebras, and Derivation Systems. Cambridge University Press, Cambridge (2008)
3. Bojadziev, G., Bojadziev, M.: Fuzzy Logic for Business, Finance and Management, 2nd edn. World Scientific Publishing Co., London (2007)
4. Bonissone, P.P., Decker, K.S.: Selecting uncertainty calculi and granularity: An experiment in trading-off precision and complexity. In: Kanal, L.N., Lemmer, J.F. (eds.) Uncertainty in Artificial Intelligence, pp. 217–247. Elsevier, Amsterdam (1986)
5. Boole, G.: The calculus of logic. Cambridge and Dublin Math. J. **III**, 183–198 (1848)
6. Donzé, L., Meier, A.: Applying fuzzy logic and fuzzy methods to marketing. In: Meier, A., Donzé, L. (eds.) Fuzzy Methods for Customer Relationship Management and Marketing, pp. 1–14. Business Science Reference, Hershey (2012)
7. Dubois, D., Prade, H.: Fuzzy Sets and Systems: Theory and Applications. Academic Press, New York (1980)
8. Fodor, J.C., Rudas, I.J.: Basics of fuzzy sets. In: Kacprzyk, J., Pedrycz, W. (eds.) Springer Handbook of Computational Intelligence, pp. 159–170. Springer, Heidelberg (2015)
9. Fukami, S., Mizumoto, M., Tanaka, K.: Some considerations on fuzzy conditional inference. Fuzzy Sets Syst. **4**, 243–273 (1980)
10. Galindo, J.: Introduction and trends to fuzzy logic and fuzzy databases. In: Galindo, J. (ed.) Handbook of Research on Fuzzy Information Processing in Databases, pp. 1–33. Science Reference, Hershey (2008)
11. Goguen, J.A.: L-fuzzy sets. J. Math. Anal. Appl. **18**, 145–174 (1967)
12. Gorzałczany, M.: Computational Intelligence Systems and Applications. Physica Verlag, Heidelberg (2002)

13. Gottwald, S.: A Treatise on Many-Valued Logics. Research Studies Press, Baldock (2001)
14. Gupta, M.M., Qi, J.: Theory of t-norms and fuzzy inference methods. Fuzzy Sets Syst. **40**, 431–450 (1991)
15. Hájek, P.: Meta Mathematics of Fuzzy Logic. Kluwer Academic publishers, Dordrecht (1998)
16. Hudec, M.: Managing fuzziness of real world in business informatics. In: 11th Conference on Strategic Management and its Support by Information Systems (SMSIS 2015), pp. 221–228, Uherské Hradiště (2015)
17. Hudec, M.: Issues in construction of linguistic summaries. In: Mesiar, R., Bacigál, T. (eds.) Proceedings of Uncertainty Modelling 2013, pp. 35–44. Slovak Technical University, Bratislava (2013)
18. Keefe, R.: Theories of Vagueness. Cambridge University Press, Cambridge (2000)
19. Klement, E.P., Mesiar, R., Pap, E.: Triangular norms: basic notions and properties. In: Klement, E.P., Mesiar, R. (eds.) Logical, Algebraic, Analytic, and Probabilistic Aspects of Triangular Norms, pp. 17–60. Elsevier, Amsterdam (2005)
20. Klement, E.P., Mesiar, R., Pap, E.: Triangular Norms. Kluwer Academic Publishers, Dordrecht (2000)
21. Klir, G., Yuan, B.: Fuzzy Sets and Fuzzy Logic: Theory and Applications. Prentice Hall, New Jersey (1995)
22. Lee, K.H.: First Course on Fuzzy Theory and Applications. Advances in Intelligent and Soft Computing, vol 27. Springer, Heidelberg (2005)
23. Liu, Y., Kerre, E.E.: An overview of fuzzy quantifiers. (I). Interpretations. Fuzzy Sets Syst. **95**, 1–21 (1998)
24. Liu, Y., Kerre, E.E.: An overview of fuzzy quantifiers. (II). Reasoning and applications. Fuzzy Sets Syst. **95**, 135–146 (1998)
25. de Luca, A., Termini, S.: A definition of a nonprobabilistic entropy in the setting of fuzzy sets theory. Inform. Control **20**, 301–312 (1972)
26. Menger, K.: Statistical metric spaces. Proc. Nat. Acad. Sci. **28**, 235–237 (1942)
27. Meier, A., Werro, N., Albrecht, M., Sarakinos, M.: Using a fuzzy classification query language for customer relationship management. In: 31th Conference on Very Large Data Bases (VLDB 2005), pp. 1089–1096, Trondheim (2005)
28. Meyer, A., Zimmermann, H.J.: Applications of fuzzy technology in business intelligence. Int. J. Comput. Commun. Control. V **I**(3), 428–441 (2011)
29. Mizumoto, M., Zimmermann, H.J.: Comparison of fuzzy reasoning methods. Fuzzy Sets Syst. **8**, 253–283 (1982)
30. Nguyen, H.T., Walker, E.A.: A First Curse in Fuzzy Logic. CRC Press, Boca Raton (1997)
31. Pavese, F.: Why should correction values be better known than the measurand true value? J. Phys. Conf. Ser. **459** (2013)
32. Pedrycz, W., Gomide, F.: An Introduction to Fuzzy Sets: Analysis and Design. The MIT Press, Cambridge MA (1998)
33. Perny, P., Roy, B.: The use of fuzzy outranking relations in preference modelling. Fuzzy Sets Syst. **49**, 33–53 (1992)
34. Perović, A., Takači, A., Škrbić, S.: Towards the formalization of fuzzy relational database queries. Acta Polytechnica Hung. **6**, 185–193 (2009)
35. Perović, A., Takači, A., Škrbić, S.: Formalising PFSQL queries using LP1/2 fuzzy logic. Math. Struct. Comput. Sci. **22**, 533–547 (2012)
36. Radojević, D.: Interpolative realization of Boolean algebra as a consistent frame for gradation and/or fuzziness. In: Nikravesh, M., Kacprzyk, J., Zadeh, L.A. (eds.) Forging new Frontiers: Fuzzy Pioneers II, Studies in Fuzziness and Soft Computing, pp. 295–317. Springer, Berlin Heidelberg (2008)
37. Rosenfield, A.: Fuzzy graphs. In: Zadeh, L.A., Fu, K.S., Tanaka, K., Shimura, M. (eds.) Fuzzy Sets and their Applications to Cognitive and Decision Processes, pp. 77–95. Academic Press, New York (1975)
38. Ruspini, E.H.: A new approach to clustering. Inf. Control **15**, 22–32 (1969)
39. Russel, B.: Vagueness. Australasian J. Psychol. Philos. **1**, 84–92 (1923)

40. Sánchez, D., Trillas, E.: Measures of fuzziness under different uses of fuzzy sets. In: Greco, S., Bouchon-Meunier, B., Coletti, G., Fedrizzi, M., Matarazzo, B., Yager, R.R. (eds.) Advances in Computational Intelligence, Part 2, vol. 298, pp. 25–34. Springer-Verlag, Heidelberg (2012)
41. Schweizer, B., Sklar, A.: Probabilistic Metric Spaces. North-Holland, Amsterdam (1983)
42. Trillas, E., Alsina, C.: On the law $p \wedge q \to r = [(p \to r) \vee (q \to r)]$ in fuzzy logic. IEEE Trans. Fuzzy Syst. **10**, 84–88 (2002)
43. Trillas, E., Cubillo, S., del Campo, C.: When QM-operators are implication functions and conditional fuzzy relations? Int. J. Intell. Syst. **15**, 647–655 (2000)
44. Viertl, R.: Fuzzy data and information systems. In: 15th WSEAS International Conference on Systems, pp. 83–85, Corfu (2011)
45. Wygralak, M.: Cardinalities of Fuzzy Sets. Studies in Fuzziness and Soft Computing, vol. 118. Springer, Berlin Heidelberg (2003)
46. Wang, H., Lee, S., Kim, J.: Quantitative comparison of similarity measure and entropy for fuzzy sets. In: Huang, R., Yang, Q., Pei, J., Gama, J., Meng, X., Li, X. (eds.) Advanced Data Mining and Applications. LNCS (LINAI), vol. 5678, pp. 688–695. Springer, Heidelberg (2009)
47. Wang, X., Ruan, D., Kerre, E.E.: Mathematics of Fuzziness Basic Issues. Studies in Fuzziness and Soft Computing, vol. 245. Springer, Berlin Heidelberg (2009)
48. Yager, R.R.: Connectives and quantifiers in fuzzy sets. Fuzzy Sets Syst. **40**, 39–75 (1991)
49. Yager, R.R.: On the measure of fuzziness and negation part I: membership in the unit interval. Int. J. Gen. Syst. **5**, 221–229 (1979)
50. Yeh, R.T., Bang, S.Y.: Fuzzy relations, fuzzy graphs and their applications to clustering analysis. In: Zadeh, L.A., Fu, K.S., Tanaka, K., Shimura, M. (eds.) Fuzzy Sets and Their Applications to Cognitive and Decision Processes, pp. 125–150. Academic Press, New York (1975)
51. Ying, M.: Implication operators in fuzzy logic. IEEE Trans. Fuzzy Syst. **10**, 88–91 (2002)
52. Zadeh, L.A.: From computing with numbers to computing with words—from manipulation of measurements to manipulation of perceptions. In: Wang, P. (ed.) Computing with Words, pp. 35–68. Wiley, New York (2001)
53. Zadeh, L.A.: Soft computing and fuzzy logic. IEEE Softw. **11**, 48–56 (1994)
54. Zadeh, L.A.: A computational approach to fuzzy quantifiers in natural languages. Comput. Math. Appl. **9**, 149–184 (1983)
55. Zadeh, L.A.: The concept of a linguistic variable and its application to approximate reasoning: part I. Inf. Sci. **8**, 199–249 (1975)
56. Zadeh, L.A.: Fuzzy sets. Inf. Control **8**, 338–353 (1965)
57. Zimmermann, H.J.: Fuzzy Set Theory—and its Applications. Kluwer Academic Publishers, Dordrecht (2001)

Chapter 2
Fuzzy Queries

Abstract The goal of database queries is to separate relevant tuples from non-relevant ones. The common way to realize such a query is to formulate a logical condition. In classical queries, we use crisp conditions to describe tuples we are looking for. According to the condition, a relational database management system returns a list of records. However, user's preferences in what should be retrieved, are often vague or imprecise. These preferences can be expressed in atomic conditions and/or between them. For example, the meaning of a query: *find municipalities with small population density and altitude about 1000 m above sea level* can be understood at the first glance. The linguistic terms clearly suggest that there is a smooth transition between acceptable and unacceptable records. This chapter is focused on the construction of fuzzy sets, the aggregations functions and the issues of fuzzy logic in queries which should not be attenuated.

2.1 From Crisp to Fuzzy Queries

In order to work with the main topic of this chapter, brief introduction to relational databases and relational algebra is desirable. Relational databases are further examined in Chap. 5.

A relational database consists of relations (tables). We should emphasize that tables and relations are synonyms. The table is a suitable representation of relation. The relation schema has the following structure [38]:

$$R(A_1 : D_1, ..., A_n : D_n), \tag{2.1}$$

where R is the name of relation, e.g. MUNICIPALITY or CUSTOMER (in order to keep notation consistent throughout the book, relations in database are written with capital letters), A_i is the i-th attribute ($i = 1, ..., n$), often called column (e.g. *unemployment rate*) and D_i is the domain of attribute A_i defining a set of all possible values which could be assigned to records (e.g. interval [0, 100] for the above mentioned attribute). In case of attributes like *sales* the domain is set of positive real numbers. A record is represented by row and often called tuple. A relation instance

© Springer International Publishing Switzerland 2016
M. Hudec, *Fuzziness in Information Systems*,
DOI 10.1007/978-3-319-42518-4_2

of a given relation schema is a set of tuples stored in a table. Hence, the term "relation instance" is abbreviated to relation or table. Each tuple $r_j (j = 1, ..., m)$ consists of the attributes' values in the following way:

$$r_j = \left\{ (d_{1j}, ..., d_{nj}) \mid (d_{1j} \in D_1, ..., d_{nj} \in D_n) \right\}, \qquad (2.2)$$

where d_{ij} is the value of the tuple r_j corresponding to the attribute A_i. The letter t is usually used in literature to express database tuple. In this book letter t is used for t-norms. In order to avoid any misunderstanding, t-norm is marked with letter t and database tuple with letter r (record) throughout the book.

To clarify the understanding of relations to level required for queries, let us show one illustrative example.

Example 2.1 The relation MUNICIPALITY(#id, name, number_of_inhabitants, area, altitude, pollution) is represented in Table 2.1.

For instance, third tuple is expressed as vector (2.2):
$r_3 = \{(3, \text{Mun3}, 810, 1030, 625, 0.20)\}$. □

A query against a collection of data stored in database provides a formal description of the tuples of interest to user posing this query [30]. The Structured Query Language (SQL) is a standard query language for relational databases [13]. SQL was initially introduced in [11]. Since then, SQL has been used in many relational databases for managing data (insert, modify, delete, retrieve). The use of SQL may be regarded as one of the major reasons for the success of relational databases in the commercial world [45].

SQL has the following basic structure:

SELECT [distinct]⟨*attributes*⟩ FROM ⟨*relations*⟩ WHERE ⟨*condition*⟩ (2.3)

In the traditional (crisp) SQL condition a tuple can either fully satisfy the intent of a query Q_c, or not. Other options do not exist. In the set theory we can express set of crisp answers in the following way:

$$A_{Q_c} = \{(r, \varphi(r)) \mid r \in R \wedge \varphi(r) = 1\}, \qquad (2.4)$$

where $\varphi(r) = 1$ indicates that the selected tuple r meets the query criterion and R states for the queried relation. Therefore, it is not necessary to write the answer as an ordered pair (tuple, matching degree).

Example 2.2 An example of crisp SQL query is:

SELECT name
FROM municipality
WHERE altitude < 250 and pollution > 20.

This query returns two municipalities (Mun 2 and Mun 3) from the relation shown in Table 2.1.

Table 2.1 Relation MUNICIPALITY in a database

#Id[a]	Name	Number_of_inhab.	Area	Altitude	Pollution
1	Mun 1	1550	2536	251	19.93
2	Mun 2	1790	7995	248	20.50
3	Mun 3	810	1030	625	0.20
4	Mun 4	5810	8030	126	60.00

[a]# represents the primary key, i.e. attribute(s) which unambiguously identify tuple

The result of the query is shown in graphical mode in Fig. 2.1. Values 20 and 250 delimit the space of retrieved data. Small squares stands for municipalities. From the graphical interpretation it is evident that two tuples (circled), although share almost the same values of both attributes, are separated (one is selected, whereas another is not).

□

If SQL is used for solving this problem, the relaxation would have to be done in the following way [12]:

SELECT name
FROM municipality
WHERE pollution $> 20 - l_1$ and altitude $< 250 + l_2$

where parameters l_1 and l_2 are used to expand the initial query condition in order to select records that almost meet the query condition. However, this approach has two disadvantages [12]. First, the meaning of the initial query is diluted (e.g. instead of $altitude < 250$ we have $altitude < 260$, if $l_1 = 10$) in order to capture adjacent tuples situated just beyond the border of the initial query. The meaning of the query is

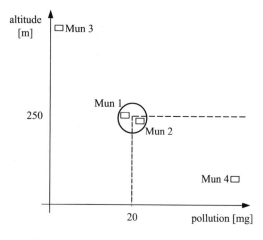

Fig. 2.1 Crisp query condition: WHERE altitude < 250 and pollution > 20

changed and the adjacent tuples satisfy the condition in the same way as initial ones. Second, a problem arises from the question: what about tuples that are very close to satisfy the new expanded query? Should we add another relaxation of query? In this way more data from the database are selected, but the initial intent of the query is lost [24].

The other option is adjusting the *select* clause in the following way:

SELECT name, altitude, (iif altitude < 250, 1, iif(altitude > 250 and altitude < 260, (260 - altitude) / 10), 0)) as matching degree

In this way, the structure of the *select* clause is a bit complicated, even though only one attribute is relaxed. If further attributes were added, then the clause would become almost illegible, e.g.

SELECT name, altitude, (iif altitude < 250, 1, iif(altitude > 250 and altitude < 260, (260 - altitude)/10),0)) as m1, pollution, iif(pollution > 20, 1, iif(pollution > 17 and pollution < 20, (pollution - 17)/3, 0)) as m2, min(m1, m2) as matching degree

If we would like to use further possibilities of fuzzy logic such as different t-norms or non-linear fuzzy sets, then rewriting the *select* clause is a complicated task which may eventuate in higher occurrence of errors.

SQL is optimized to query relational databases. In this chapter the core of SQL remains intact and the extension is done to fuzzify the imprecise conditions. Adding flexibility to SQL meets requirements for flexible queries and increases effectiveness of the whole querying process.

The main reason for using fuzzy set theory to make querying more flexible is discussed in [15] and advocated in [27]. Firstly, fuzzy sets provide a better description of data requested by user. For example, the meaning of a query: *select customers having high turnover and short payment delay* can be understood at the first glance. Secondly, linguistic terms are clearly suggesting that there is a smooth transition between acceptable and unacceptable tuples. In fuzzy queries, some tuples definitely match the condition, some certainly not and some match to a certain degree.

Similarly to SQL query structure (2.3), the basic structure of fuzzy query is the following [5]:

SELECT [distinct]⟨*attributes*⟩ FROM ⟨*relations*⟩ WHERE ⟨*fuzzy condition*⟩
$$(2.5)$$

Hence, the fuzzy query returns a fuzzy relation (a subrelation of the initial database relation) consisting of set of tuples that satisfy the fuzzy condition and respective matching degrees. The set of answers to fuzzy query A_{Q_f} could be written in the following way:

$$A_{Q_f} = \{(r, \mu(r)) \mid r \in R \wedge \mu(r) > 0\} \qquad (2.6)$$

In fuzzy queries we distinguish preferences inside atomic conditions and between them. The former is expressed by constructing fuzzy sets which correspond to users' needs; the latter is realized by aggregations.

Presumably, the first attempt to fuzzify SQL-like queries is [42]. It fuzzifies the *where* clause in order to make possible to use vague terms similar to natural language. The aggregation was realized by the minimum t-norm (1.47) for the *and* connective and by the maximum s-norm (1.59) for the *or* connective. It was the inspiration for practical realizations of fuzzy queries. One of the first practical realizations of flexible queries is FQUERY, an add-in that extends the MS Access's querying capabilities with the linguistic terms inside the *where* clause [28]. SQLf [5, 6] is a more comprehensive fuzzy extension of SQL queries. SQLf extends SQL by incorporating fuzzy predicates in the *where* clause as well as supports, among others, subqueries inside the fuzzified *where* clause and fuzzy *joins*. The Fuzzy Query Language (FQL) [47] extends SQL queries with fuzzy condition inside the *where* clause in the usual way and adds other two clauses which provide additional functionality: *weight* and *threshold*.

Flexible querying is still an active field for research either in adding further flexibility such as the fuzzification of the *group by* clause [9], conversion from fuzzy to crisp queries [24], dealing with the empty and overabundant answer problems [7, 40] and user-friendly graphical interface [39]. An exhaustive source of flexible preference querying is [34].

2.2 Construction of Fuzzy Sets for Flexible Conditions

Generally, there are two main aspects for constructing fuzzy sets. In the first aspect, users define the parameters of each fuzzy set according to their opinion. This way gives them freedom to choose parameters (a, b, m, m_1, m_2) depicted in Figures (1.5, 1.6, 1.7, 1.8, 1.9 and 1.10). Nevertheless, in this approach users are asked to set more crisp values than in classical query (e.g. two crisp values to clarify the meaning of term *small* (1.22), whereas in the classical query user assigns only one value, e.g. *attribute A* $< a$). This problem can be mitigated by adjusting fuzzy sets parameters by moving sliders (on interfaces) over the domain of attribute to set ideal and acceptable values [39] rather than filling input fields with crisp numbers, for example.

Let D_{min} and D_{max} be the lowest and the highest possible domain values of a numeric attribute A, i.e. $Dom(A) = [D_{min}, D_{max}]$ and L and H be the lowest and the highest values in the current content of a database, respectively. The domains of attributes should be defined during the database design process in a way that all theoretically possible values can be stored. For the attribute describing the daily frequency of a measured phenomenon during the year (e.g. *number of days with temperature below* $0\,°C$), the domain is the $[0, 365]$ interval of integers. In practice, the real values could be far from the D_{min} and D_{max} values; that is, $[L, H] \subseteq [D_{min}, D_{max}]$. This means that only part of the domain contains data. For the attribute describing the number of inhabitants, the domain is theoretically the whole set of natural numbers ($\mathbb{N}$), but stored values are far from the infinity (∞). If user is not aware of these facts, the query might easily end up as empty or overabundant. This fact should be considered in defining not only query conditions, but also inference rules and data summarizing sentences.

The second aspect deals with the dynamic modelling fuzzy sets parameters over the domains of attributes. In the first step values of L and H are retrieved from a database. These parameters are used to create fuzzy sets for attributes included in the query condition.

The modelling of fuzzy sets parameters depends on the type and purpose of the fuzzy query (dependable or undependable atomic conditions). In this section several approaches are examined. In the next sections suitable ways for particular tasks are discussed.

If collected data are more or less uniformly distributed in the domain, then the uniform domain covering method [44] can be applied. Otherwise, the statistical mean-based method [44] or the logarithmic transformation of respective domains [25] can be used.

An appropriate method could be chosen by users or mined from the data. In the first case, users could rely on their knowledge, common sense, attainments about examined entities (e.g. territorial units or customers), and knowledge of physical laws. Let us demonstrate this aspect on the municipal database. For attributes like the number of inhabitants, the choice may be the logarithmic transformation or statistical mean-based algorithm, because the database contains a few big municipalities and majority of smaller ones. In the case of attributes like water or gas consumption per inhabitant, users could assume that the uniform domain covering method is an appropriate option.

In the second case, per-computation (performing an initial computation before run the main task) of data summary on attributes is used to reveal the information about distribution of data. In this way the single scan of database provides relevant information about relative cardinality [40].

In the book we use the uniform domain covering method. The parameters for the linguistic terms are created by calculating the cores and slopes in the following way [44]:

$$\varepsilon = \frac{1}{8}(H - L) \tag{2.7}$$

$$\theta = \frac{1}{4}(H - L) \tag{2.8}$$

when the linguistic variable is divided into three terms. Thereafter, it is easy to calculate required parameters shown in Fig. 2.2 for a particular fuzzy set. If it is a requirement for more fuzzy sets (e.g. five sets: very small, small, medium, high, very high, Fig. 1.17) these sets can be straightforwardly constructed adjusting $\varepsilon = \frac{1}{14}(H - L)$ and $\theta = \frac{1}{7}(H - L)$.

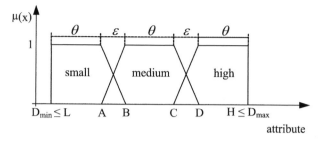

Fig. 2.2 Linguistic variable uniformly distributed over an attribute domain by three fuzzy sets

Uncertainty of belonging to a set is expressed by ε. The flat segment defined by θ express clear belonging to a set. If uncertainty decreases, the value of ε decreases. In the extreme situation, i.e. $\varepsilon = 0$ we got the classical partitioning of the attribute's domain.

Example 2.3 An institution analysing usage of water is interested to find municipalities which have small water consumption per inhabitants in cm^3. For this task the granularity to three fuzzy sets should be sufficient.

Let a simple SQL:

SELECT max(WaterConsumpt) as L, min(WaterConsumpt) as H
FROM municipality

reveals: $L = 0.63$ and $H = 389.22$. Straightforwardly, applying (2.7) and (2.8) only parameters for the fuzzy set *small* are calculated, that is, $A = 97.7775$ and $B = 146.35125$.

Therefore, we got $m_2 = A = 97.7775$ and $b = B = 146.35125$ in order to keep the notation in accordance to the notation in Sect. 1.2.2. Afterwards, user can either modify these parameters or run a query with the suggested ones. In case of the latter, the *where* clause is:

WHERE WaterConsumpt < 146.35125.

This condition ensures that all tuples which fully or partially meet the condition are selected for the second step: calculating matching degrees. □

Another approach for constructing fuzzy sets is presented in [27] to support fuzzy queries such as *select municipalities where number of warm days is much lower than number of days with snow coverage*. The intervals $[L, H]$ of each database attribute are transformed into the $[-10, 10]$ interval of real numbers. In the case of condition like *altitude above sea level is small and number of inhabitants is high*, attributes' values can be also transformed into the $[-10, 10]$ interval, if the data distribution is more or less uniform. In this case, fuzzy sets *small, medium, high* can be created using $\varepsilon = 3.3$ and $\theta = 6.6$.

These linear shapes of membership functions, although very suitable for fuzzy queries, constitute only a small subset of the all possible shapes of membership functions. Discussion about other ones can be found in, e.g. [17, 32].

Both aspects of constructing fuzzy sets parameters can be merged. In the first step, parameters of fuzzy sets are calculated from the current content of database. In the second step, users are able to modify these parameters, if they are not satisfied with the suggested ones. Obviously, the user can obtain tentative information about the stored data before running a query. The querying process could become more tedious, but on the other hand, it might save computation time from running queries which return empty sets. In case when suggested parameters are far from the user's expectations, user has two options: (i) to quit the query process knowing that there is no data that satisfy the query condition; (ii) to accept or slightly modify the suggested parameters. The assistance for constructing fuzzy sets parameters should be optional for users.

The topic of construction of fuzzy sets is covered by vast literature, mainly in fuzzy systems. We should be careful when we consider constructing fuzzy sets for queries, due to the following reasons [20]:

- In fuzzy inference systems fuzzy sets should cover the whole domains of attributes in order to properly control or classify all possible occurrences of input attributes. On the contrary, queries select a subset of data which might be relevant for users.
- If the goal is to develop an easy to use and less demanding tool (a web application for the broad audience, for example), then the fuzzification process should be as simple as possible. It means that, sophisticated approaches like neural networks or genetic algorithms should be avoided.
- In queries, where answer to the first part of query influences adjusting the second part of query we have to rapidly and efficiently construct fuzzy sets for dependable attributes during the querying process.

In addition, the fuzzification step can be improved by mining parameters from the recorded history, if application keeps user's preferences. During the next attempt for similar query, the application offers parameters of fuzzy sets from the recorded user's history. The second option is recommending parameters according to the parameters used by similar users, e.g. if a 22-year-old student from the Eastern Europe searches for a hotel, then the procedure could approximately guess meanings of terms related to the price and room size from other similar students. The main drawback is keeping the user's history and building recommendations. On the other hand, this way is more tailored to users and may attract them. Anyway, finding suitable options for each application should be considered.

2.3   Converting Fuzzy Conditions to SQL Ones

If we wish to effectively select tuples, flexible conditions should be converted into SQL ones, because SQL is optimized for efficiently querying relational databases. This task is explained on the approach based on the Generalized Logical Condition

(GLC) [24]. This solution contains the usual steps for fuzzy querying: (i) converting fuzzy conditions to SQL ones; (ii) connecting to database, selecting all candidates (tuples which have membership degree greater than zero) and releasing a database connection; (iii) calculating satisfaction degree for each tuple to each atomic condition; (iv) calculating overall satisfaction degree (often called matching degree). The detailed explanation of this approach can be found in [23].

The GLC has the following structure:

$$where \ \otimes_{i=1}^{n} A_i \circ L_i, \tag{2.9}$$

where n denotes the number of attributes in fuzzy condition of a query,

$$\otimes = \begin{cases} and \\ or \end{cases} \text{ where } and \text{ and } or \text{ are fuzzy logical operators and}$$

$$A_i \circ L_i = \begin{cases} A_i > a, & \text{for condition } high \\ A_i < b, & \text{for condition } small \\ A_i > a \text{ and } A_i < b, & \text{for condition } medium \end{cases}$$

where A_i is i-th attribute included in the condition, parameters a and b delimit supports of respective fuzzy sets explained in Sect. 1.2.2.

When the compound condition contains several atomic ones connected with the *and* operator, then tuple meets partially or fully the overall condition only, if values of all attributes belong to supports of the respective fuzzy sets.

If a *where* clause contains fuzzy as well as classical conditions, classical ones could be easy added to the *where* clause (2.9) in the following way:

$$where \ \otimes_{i=1}^{n} A_i \circ L_i \text{ [and/or] } A_e > e \text{ [and/or] } A_f = \text{"string"} \dots \tag{2.10}$$

Example 2.4 In this example municipalities with altitude about 900 m above sea, high number of beds in accommodation facilities and small population density are sought.

In the first step each linguistic term is expressed by fuzzy set. Altitude about 900 m above sea is represented by the trapezoidal fuzzy set (1.20) with parameters: $a = 850$, $m_1 = 875, m_2 = 925$ and $b = 950$. High number of beds in accommodation facilities is expressed by the R fuzzy set (1.23) with parameters $a = 450$ and $m = 550$. Finally, small population density is described by the L fuzzy set (1.22) with parameters $m = 120$ and $b = 135$.

Now, we have all information for converting fuzzy into SQL query. The SQL query is:

SELECT name
FROM municipality
WHERE (altitude > 850 and altitude < 950) and beds_accommodation > 450 and population_density < 135.   □

When we have selected all tuples, we can continue with calculating their respective matching degrees.

2.4   Calculation of Matching Degrees

Query conditions usually consist of more than one atomic condition merged by logical connectives. Generally, we can divide this topic into two main categories.

- The simpler category is aggregation of query conditions which have the same significance for users and are independent, i.e. order of the execution is irrelevant (commutative atomic conditions).
- The more complex category are queries where elementary conditions have different relevance (commutative and non-commutative queries).
  Commutative queries can be solved by fuzzy implications or Ordered Weighted Averaging Operator (OWA).
  Non-commutative queries have several structures. Bipolar queries merge constraints (have to be satisfied) and wishes (is nice if are satisfied) or negative and positive judgements, respectively. Another example is non-commutative queries containing only constraints where answer of the first atomic condition influences answer to the second one.

In the next several subsections these categories are examined.

2.4.1   Independent Conditions Aggregated by the "And" Operator

This is the simplest form of aggregation. The t-norm functions (Sect. 1.3.1) are used as *and* logical operator. It is well known that different t-norms produce different matching degrees. Naturally, the following question has arisen: which t-norm is the most suitable one?

Let us recall the well-known fact that only the minimum t-norm (1.47) is an idempotent one. This fact is one of the reasons for the wide use of this operator. However, the minimum t-norm has a limitation. Only minimal value of all atomic conditions is considered, that is, other atomic conditions do not influence the solution. The product t-norm (1.48) takes into account membership degrees of all elementary conditions. Therefore, this t-norm distinguishes records which have the same value of the lowest membership degree of atomic conditions and different values of the satisfaction degree of other atomic conditions, but produces lower matching than the minimum t-norm. The Łukasiewicz t-norm (1.49) produces membership degree greater than 0 only for tuples which significantly satisfy the condition, i.e.:

$$\sum_{i=1}^{n} \mu_{Ai}(r) - (n - 1) > 0$$

because this t-norm is a nilpotent one. The last basic t-norm is drastic product (1.50), which is not suitable due to non-continuity and high restrictiveness.

Table 2.2 Matching degrees calculated by different t-norms

Tuple	μ_{P_1}	μ_{P_2}	Min(1.47)	Prod(1.48)	Łuk(1.49)	Drast(1.50)
r1	0.21	0.32	0.21	0.067	0.00	0
r2	0.20	0.96	0.20	0.192	0.16	0
r3	0.95	0.95	0.95	0.905	0.9	0

Let us have three records which satisfy the first atomic condition (predicate) (P_1) and the second atomic condition (P_2) as is shown in Table 2.2.

It is obvious that tuple $r3$ is the best option. But, drastic product calculates matching equal to 0. Let us now focus on tuples $r1$ and $r2$. The minimum t-norm prefers the tuple $r1$. But, this could contradict the human reasoning when selecting the best tuple: although the tuple $r1$ is only slightly better in the first atomic condition and notably worse according to the second one, it is the preferred one. The product t-norm prefers $r2$ with the membership degree lower than the values for both atomic conditions. The Łukasiewicz t-norm calculates membership degrees greater than 0 for the tuple $r2$ because $r1$ does not significantly satisfy both atomic conditions.

On the other hand, if the decision depends on the matching degree, e.g. percentage of the financial support or discount, then using non-idempotent t-norms could contradict with the usual reasoning. If tuple satisfies all atomic conditions with degree of 0.5, then it is expectable that the percentage of support is 50 % (0.5). In case of product t-norm the matching degree significantly decreases, when number of atomic conditions increases. Although product t-norm is not a nilpotent one, it converges to 0, when large number of atomic predicates exists, that is, $\lim_{n \to \infty} \prod_{i=1}^{n} \mu_{Ai}(x) = 0$. This fact is known as the limit property of t-norms [31]. On the other hand, minimum t-norm converges to minimal value of atomic predicates.

When we want to select only records which significantly meet the compound predicate (overall query condition) and avoid issue discussed in the paragraph above, then the α-cut may be the solution. But care should be taken, when not only minimum, but also sum of atomic conditions is relevant. The nilpotent minimum t-norm (1.54) may be the option for such tasks. Let us have two records which satisfy the first atomic condition (P_1) and the second atomic condition (P_2) as is shown in Table 2.3.

According to minimum t-norm tuple $r2$ is preferred. But, tuple $r1$ dominates in the first atomic condition and is slightly worse in the second one. In order to select

Table 2.3 Matching degrees calculated by minimum and nilpotent minimum t-norms

Tuple	μ_{P_1}	μ_{P_2}	Min(1.47)	Nilpotent min(1.54)
r1	0.80	0.30	0.30	0.30
r2	0.40	0.35	0.35	0

only tuples which significantly meet the compound query condition, threshold can be used. Let us have threshold 0.33. By minimum t-norm $r1$ is excluded. Contrary, nilpotent minimum t-norm prefers tuple $r1$ because $\mu_{P_1} + \mu_{P_2}$ is higher than 1 and therefore meets threshold value of 1 of this t-norm. To conclude this observation, query matching degrees of tuples which pass filtre of nilpotent minimum t-norm, are calculated in the same way as by minimum t-norm.

Contrary to the crisp conjunction, the fuzzy conjunction can be expressed by variety of t-norms. This provides a benefit because we can model a large scale of users' requirements. Nevertheless, we should carefully decide which t-norm is the most suitable in order to avoid inappropriate solutions. Hence, developers of information systems and databases should be familiar with the fuzzy set and fuzzy logic theory.

To summarize, when the most restrictive matching degree of atomic condition is required, then the minimum t-norm is the appropriate solution. Furthermore, when it is desirable that the sum of atomic predicates significantly contributes to solution, then nilpotent minimum t-norm is an option. When users consider satisfaction degrees of each atomic condition, then the product t-norm is the choice. In addition, if user wishes to see in the resulting relation only tuples which significantly meet all atomic conditions, then the Łukasiewicz t-norm, or applying α-cut on the result obtained by product t-norm are the suitable choices.

Flexible queries can be straightforwardly adjusted for searching similar entities [21] to the existing or ideal one. In this type of fuzzy query membership functions are limited to triangular ones (Fig. 1.5), because support (1.9) should be limited and membership degree equal to 1 should hold only for tuples having the same value of analysed attribute as the reference tuple. The *and* connective should not be expressed by minimum or nilpotent minimum t-norm.

If atomic conditions are merged by the *or* logical operator, the s-norms are applied. This kind of queries is not further examined, but the duality principle between t-norms and s-norms discussed in Sect. 1.3.3 helps in searching for the suitable s-norm.

Design of interfaces for commutative queries, searching for similar tuples and related discussions are in Appendix A.1.

2.4.2 Fuzzy Preferences Among Atomic Query Conditions

The aim of preferences among atomic conditions is to distinguish more important conditions from less important ones. In everyday tasks people rarely give the same priority to all attributes. As an example let us take the query: *select young and highly productive employees where age is more important than productivity*.

In order to solve such a query, weights $w_i \in [0, 1]$ can be associated with atomic conditions. Two types of weights can be applied [49]: static and dynamic. From the names it is obvious that the static weights are fixed, known in advance and unchangeable during query processing, whereas for dynamic weights neither their values nor association to criteria are fixed a priori.

The idea for calculating the matching degree of atomic conditions P_i according to an importance weight w_i by fuzzy implication has the following form [49]:

$$\mu(P_i^*, r) = (w_i \Rightarrow \mu(P_i, r)), \tag{2.11}$$

where $\Rightarrow$ represents a fuzzy implication (weight implies or influences the solution). In order to be meaningful, weights should satisfy several requirements [15]:

- if $w_i = 0$, then the result should be such as P_i does not exist or does not have any influence on the solution
- if $w_i = 1$, then P_i fully influences the solution
- weights should be assigned to each atomic condition, whereas at least one weight should have value of 1 for the most important attribute(s): $\max_i(w_i) = 1, i = 1 \dots n$

By applying these requirements, it is easy to conclude that the regular implications (S, Q, R—Sect. 1.3.4) such as Kleene-Dienes, Gödel and Goguen match the requirements.

Example 2.5 In this example we check whether the Kleene-Dienes implication (1.63) is suitable. Consider the overall query condition consisted of several atomic predicates connected with the *and* operator $\wedge_{i=1}^n P_i$.

For very low importance of the P_i (w_i is close or equal to 0), the satisfaction of elementary condition P_i has a very low influence on the query satisfaction, because: $w_i \rightarrow 0 \Rightarrow \mu(P_i^*, r) \rightarrow 1$, where arrow means approaching the value of. In the extreme situation (no importance at all, $w_i = 0$), $\mu(P_i^*, r) = 1$. The value of 1 is the neutral element in the conjunction.

For maximal importance of the P_i, w_i is equal to 1. The satisfaction of P_i is essential for fulfilling the overall query, because $w_i \rightarrow 1 \Rightarrow \mu(P_i^*, r) \rightarrow \mu(P_i, r)$.

In the same way it is possible to prove that the other regular implications are suitable. □

Using the Kleene-Dienes implication, the following query condition for the conjunction is achieved:

$$\mu(r) = \min_{i=1,\dots,n} (\max_{i=1,\dots,n}(\mu(P_i, r), 1 - w_i)) \tag{2.12}$$

if the minimum function is used as a t-norm. Other t-norms can be also used. Furthermore, the Eq. (2.12) corresponds to the definition of the weighted conjunction operator introduced in [16].

For the Gödel implication (1.67) Eq. (2.11) yields:

$$\mu(P_i^*, r) = \begin{cases} 1 & \text{for } \mu(P_i, r) \geq w_i \\ \mu(P_i, r) & \text{for } \mu(P_i, r) < w_i \end{cases} \tag{2.13}$$

The weight w_i is treated as a threshold. If predicate P_i is satisfied to a degree greater or equal than this threshold, then the weighted condition is considered to be fully satisfied. Otherwise it equals to matching degree of P_i. Furthermore, it is easy to prove that when $w_i = 0$, regardless of value of $\mu(P_i, r)$ the answer participates in the conjunction with value 1 (neutral element).

Goguen implication (1.68) is another threshold-type interpretation, but the under-satisfaction of the condition is treated in a more continuous way:

$$\mu(P_i^*, r) = \begin{cases} 1 & \text{for } \mu(P_i, r) \geq w_i \\ \frac{\mu(P_i, r)}{w_i} & \text{for } \mu(P_i, r) < w_i \end{cases} \tag{2.14}$$

Besides these implication functions, in some applications it is common to describe implication by t-norms [18]. Let us see whether this assumption holds here. For the so-called Mamdani implication (minimum t-norm), the proof of unsuitability of this implication is simple. For no importance of P_i ($\mu(P_i^*, r) = \min(w_i, \mu(P_i, r)) = 0$) the overall query satisfaction will be 0, because value of 0 annihilates truth values of other atomic conditions in conjunction. It implies that the requirement: *if $w_i = 0$, then the result should be such as if P_i does not exist* or does not have influence, is not satisfied for the Mamdani implication.

This result was expected. But the goal of this short discussion was to emphasize that the "simplified implication" used in fuzzy reasoning does not work here.

Example 2.6 Let us have two atomic predicates: P_1 having high importance ($w_1 = 1$) and P_2 having lower importance (e.g. $w_2 = 0.5$). Tuples and matching degrees are shown in Table 2.4. The second and third column shows matching degrees before applying Kleene-Dienes implication and next two after applying this implication.

Without applying preferences tuples $r1$ and $r2$ have the same matching degree regardless of used t-norm (commutativity axiom). But, when predicate P_1 is more preferred, then its higher membership degree should be reflected in the solution. □

An example of interface adjusted to preferences is illustrated in Appendix A.1.

Another way for realization of preferences can be found in [47], where the authors suggest not only crisp values, but also fuzzy sets to describe the importance value in the additional *weight* clause. The importance weights can be crisp numbers from the [0, 1] interval or fuzzy sets defined beforehand and stored in a separated database table. The benefit for users is in the possibility to select the importance defined by linguistic terms. The linguistic variable *Preferences weight* may consist of several terms such as no importance, very low, low, medium, etc. Weights can be defined in a similar way as relative quantifiers (Sect. 1.5), because domain is the unit interval.

Table 2.4 Preferences calculated by Kleene-Dienes implication

Tuple	$\mu_{P1}(r)$	$\mu_{P2}(r)$	$\mu_{P1^*}(r)$	$\mu_{P2^*}(r)$	Matching degree (2.12)
r1	0.9	0.4	0.9	0.5	0.5
r2	0.4	0.9	0.4	0.9	0.4

2.4.3 Answer to the Second Atomic Condition Depends on the Answer to the First One

It may happen in everyday tasks that the answer to the first question influences answer to the second one. This occurs in queries, where conditions are not independent, contrary to the cases explained in Sects. 2.4.1 and 2.4.2. In this case, the second atomic condition is relative to the first one.

Two gradual conditions are combined in such a way that the second condition is applied on a subset of database rows, already limited by the first one. An example of such a query is: *select companies with small number of employees* (P_1) *among companies with high export* (P_2). The condition P_2 (consisted of atomic or compound condition) is defined a priori and the condition P_1 is defined in a relative manner of satisfying the condition P_2. If we permute predicates, the query is: *select companies with high export* (P_1) *among companies with small number of employees* (P_2). Therefore, the result may be different.

This class of queries require focus on a limited subset of an attribute domain instead of the whole domain (or to be more precise, the subset of the current content of a database).

To solve such tasks efficiently, fuzzy aggregation operator called *among* is defined [44]:

$$\mu_{P_1 among P_2} = \min(\mu_{P_1/P_2}(x), \mu_{P_2}(x)), \tag{2.15}$$

where μ_{P_2} is the membership function defining fulfillment of the independent predicate and μ_{P_1/P_2} is the membership function of the dependent predicate relative to the independent one. The former is constructed directly from data or by user, whereas the latter represents a transformation of the initial membership function μ_{P_1} affected by the independent predicate.

The construction of the μ_{P_2} can be realized by any method mentioned in Sect. 2.2. The construction of the μ_{P_1/P_2} is realized by the transformation f in the following way [44]:

$$\mu_{P_1/P_2} = f \circ \mu_{P_1}, \tag{2.16}$$

where f is the transformation between initial domain $[l, h]$ and reduced one $[l', h']$

$$[l', h'] \mapsto f : [l, h] \tag{2.17}$$

$$f(x) = l + \frac{h - l}{h' - l'}(x - l') \tag{2.18}$$

Hence, the *among* operator is calculated as (2.15)–(2.18):

$$\mu_{P_1 among P_2}(x) = \min(\mu_{P_1}(l + \frac{h-l}{h'-l'}(x - l')), \mu_{P_2}(x)) \qquad (2.19)$$

Another option for constructing compressed membership function (fuzzy set) is discussed in [22]. However, this approach requires one scan and two queries. In the first step entities satisfying the condition P_2 are selected. Tuples selected by P_2 create a subrelation of all tuples. It causes the compression of initial domain, i.e. $[L_{P1-compr}, H_{P1-compr}] \subseteq [L_{P1}, H_{P1}]$ of the dependent attribute P_1, where L_{P1} and H_{P1} represent the lowest and the highest value of dependent attribute in the whole relation, respectively.

In the second step, a database scan retrieves values of $L_{P1-compr}$ and $H_{P1-compr}$. In the third step, the fuzzy set describing dependable condition is created on the subdomain $[L_{P1-compr}, H_{P1-compr}]$ by the uniform domain covering method [44]. Even if user can define parameters for the membership function μ_{P_2} without suggestion from the current database content, defining the membership function for μ_{P_1/P_2} on the interval $[L_{P1-compr}, H_{P1-compr}]$ depends on the selected tuples in the first step. Therefore, this step should be automatized.

Finally in the last step, the overall query matching degree is calculated by (2.15). This procedure is shown in Fig. 2.3.

Example 2.7 A small survey is conducted to find expensive books among books with small number of pages. Books from a bookshop are shown in Table 2.5. Independent predicate *number of pages is small* is expressed as L fuzzy set (Fig. 1.8) with parameters $m = 200$ and $b = 250$. This condition returns five books, which fully or partially match independent predicate, shown in Table 2.6.

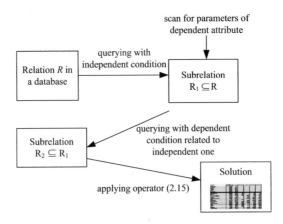

Fig. 2.3 The procedure for calculating matching degree when dependable condition is not defined a priori but in a relative manner

Table 2.5 Books in a bookshop

Book	Pages	Price
Book 1	420	60
Book 2	500	25
Book 3	290	50
Book 4	210	36
Book 5	100	45
Book 6	120	38
Book 7	225	10
Book 8	240	50
Book 9	310	70
Book 10	300	30

Table 2.6 Selected books from Table 2.5 by independent condition

Book	Pages	Price
Book 4	210	36
Book 5	100	45
Book 6	120	38
Book 7	225	10
Book 8	240	50

It is now a straightforward task to detect in the Table 2.6 the smallest value ($L_{compr} = 10$) and the highest value ($H_{compr} = 50$) in the subdomain of price attribute.

In the next step, by the uniform domain covering method parameters of fuzzy set *price is high* are calculated: $\theta = \frac{1}{4}(H_{compr} - L_{compr}) = 10$, $\varepsilon = \frac{1}{2}\theta = 5$ and then the *price is high* set is expressed as R fuzzy set with parameters $a = C = 35$ and $m_1 = D = 40$ (Fig. 2.2). Finally, by (2.15) the result is shown in Table 2.7.

Clearly, if permutation of attributes is realized, then the answer may be different. If a survey searches for small paged books among expensive ones, then the high price is analysed on the whole domain and small number of pages on the reduced subdomain. By the same procedure, the answer consists of two books: *Book 3* and *Book 8* with the matching degree of 0.33 each. □

Table 2.7 Solution by the *among* connective

Book	Pages	$\mu_{Pages}(r)$	Price	$\mu_{Price/Pages}(r)$	Among
Book 5	100	1	45	1	1
Book 6	120	1	38	0.6	0.6
Book 4	210	0.8	36	0.2	0.2
Book 8	240	0.2	50	1	0.2

The commutativity and monotonicity properties are not satisfied, whereas the associativity and existence of unit element are [44]. The associativity property allows creating encapsulated dependencies, e.g. P_1 *among* P_2 *among* P_3. It is crucial for users to properly define parentheses. The condition could be for example, *(small polluted municipalities among high sized) among high unemployed* as well as *small polluted municipalities among (high sized among high unemployed)*.

Predicates P_1, P_2 could be atomic conditions such as in Examples 2.7 and A.4 or compound ones, for instance *select municipalities with (small altitude and high pollution)* (P_1) *among municipalities with (high unemployment and high population density)* (P_2). The same discussion related to choice of suitable conjunction in Sect. 2.4.1 holds here for calculating matching degrees inside the dependent and independent compound predicates.

The interface from cases examined in Appendix A.1 has been adjusted to this kind of tasks in Appendix A.2.

2.4.4 Constraints and Wishes

Not all requirements for data can be represented as constraints. When people express their requirements they could have in mind wishes as well. A suitable example is *find hotel which has low price and possibly short distance to point M*. The budget is the limitation; we cannot afford something beyond our budget. The distance is a wish. We would prefer shorter walk, if possible. These tasks can be solved either by bipolar queries (where conditions are fully independent), or by non-commutative operators keeping the wishes and constraints together. In the book the focus is put on the latter, due to simpler way for creating applications for non-expert users. In order to mention both aspects, bipolar queries are discussed as well.

We can formally write query as [48]: *find tuples satisfying N and if possible P*, where N denotes negative preference and P describes positive preference. Answer to a bipolar query is written in the following way:

$$A_{Q_{fbp}} = \{r \mid N(r) \ and \ possibly \ P(r)\} \tag{2.20}$$

It is evident that this kind of queries is not commutative. Query *short distance and possible low price* has a different semantic meaning (budget is not a problem but distance is).

This type of queries cannot be solved by *and* operator, weights or averaging aggregations. Let us consider the low price and short distance atomic predicates from the aforementioned example. When both predicates are either fully satisfied, or fully rejected the answer is clear. When low price is fully unsatisfied, then the answer should be zero, regardless of the satisfaction degree of short distance. But, when low price predicate is fully satisfied and short distance predicate fully unsatisfied, the answer should be lower than 1, but greater than 0. Weight attached to the short distance predicate is not able to provide solution.

2.4.4.1 Bipolar Queries

The first attempt to solve bipolar queries (2.20) was realized for crisp ones. In this classical approach the condition for tuple r is expressed as [33]:

$$N(r) \ and \ possibly \ P(r) = N(r) \wedge \exists s(N(s) \wedge P(s)) \Rightarrow P(r) \qquad (2.21)$$

In the first step tuples satisfying N are selected from a database. This step ensures that tuples which do not satisfy N are not considered. If no tuple meets N, then answer to bipolar query is empty. In the second step tuple r is preferred, if no other tuples satisfies P or tuple r in the best way satisfies P. The approach: first to select by using N, then to order by using P cannot be directly applied when satisfaction is a matter of degree [14, 48].

The crisp structure of the bipolar query (2.21) is expressed in fuzzy terms in the following way [48]:

$$A_{Q_{(N,P,R)}} = \{(r, \mu(r)) \mid min(N(r), max(1 - max_{s \in R} min(N(s) \wedge P(s)), P(r)))\}, \qquad (2.22)$$

where (N, P, R) means answer to N and P against the set of tuples R. This equation is the generalization of the (2.21) from crisp to fuzzy logic. Quantifier $\exists$ is modelled by the *maximum* operator. The implication is characterized by the Kleene-Dienes one (1.63). Minimum t-norm, maximum s-norm and standard negation form a triplet characteristic by the fact that minimum t-norm is dual to maximum s-norm when the standard negation is applied. Other triplets like (Łukasiewicz t-norm and s-norm, standard negation) or (product, probabilistic sum, standard negation) can be also applied, if reinforcement effect is needed (different functions provide different matching degrees). Influences of different functions for quantifier, implication, t-norm and s-norm on the solution are discussed in [48].

Formula (2.22) expresses the global interpretation of the term *and if possible*, i.e. checking whether the constraint is satisfied by at least one tuple from the dataset considered and comparing with other tuples. On the other side, in the local interpretation (Sect. 2.4.4.2) satisfying the constraint provides a benefit to the tuple, but there is no need to compare with the other tuples.

Example 2.8 Let us have four houses satisfying N (low price) and P (short distance) with degrees depicted in Table 2.8. The matching degree to query condition obtained by the bipolar *and if possible* operator (2.22) is in the last column. The high satisfaction of wish is not as important as high satisfaction of constraint. Furthermore, influences of other tuples are considered. In the first step, tuples satisfying N are considered. This step ensures that tuples which do not satisfy N are excluded. In the second step tuple r is preferred, if no other tuples better meet P or tuple r satisfies P. Hence, the best house is $Ho1$. □

Table 2.8 Bipolar *and if possible operator*

House	$N(r)$	$P(r)$	$A_{Q_{(N,P,R)}}(r)$ (2.22)
Ho 1	0.8	0.5	0.7
Ho 2	0.8	0.3	0.5
Ho 3	0.2	0.7	0.2
Ho 4	0.1	0.9	0.1

The second approach for managing bipolarity is based on the possibility theory [14]. The answer is measured on bipolar scales: either on one scale, where the middle point is neutral and ends bear extreme positive or extreme negative values; or on two scales, one scale measures the positive and the other one negative preferences. The third approach for bipolar queries is based on the lexicographic ordering [3]. In this approach degrees for N and P are evaluated separately, i.e. no aggregation between constraint and wish is performed.

2.4.4.2 Non-commutative Operators

Four non-commutative operators have been introduced in [4]:

- P_1 *and if possible* P_2—relaxation of conjunction
- P_1 *or else* P_2—intensification of disjunction
- P_1 *all the more as* P_2
- P_1 *all the less as* P_2

In this book first two operators are considered. Instead of bipolarity, this approach is focused on relaxing conjunction and intensification disjunction, respectively. These operators are also called asymmetric conjunction and asymmetric disjunction correspondingly.

Bosc and Pivert [4] created the following six axioms in order to formally write *and if possible* operator:

- is less restrictive than the *and* operator (P_1 *and* P_2), i.e. $\alpha(\mu_{P_1}, \mu_{P_2}) \geq \min(\mu_{P_1}, \mu_{P_2})$
- is more drastic than only constraint (P_1) appears, i.e. $\alpha(\mu_{P_1}, \mu_{P_2}) \leq \mu_{P_1}$;
- is increasing in constraints argument, i.e. $a \geq b \Rightarrow \alpha(a, c) \geq \alpha(b, c)$;
- is increasing in wishes argument, i.e. $b \geq c \Rightarrow \alpha(a, b) \geq \alpha(a, c)$;
- has asymmetric behaviour, i.e. $\alpha(\mu_{P_1}, \mu_{P_2}) \neq \alpha(\mu_{P_2}, \mu_{P_1})$;
- P_1 *and if possible* P_2 is equivalent to P_1 *and if possible* (P_1 *and* P_2), i.e. $\alpha(\mu_{P_1}, \mu_{P_2}) = \alpha(\mu_{P_1}, \min(\mu_{P_1}, \mu_{P_2}))$.

Hence, function h of the structure:

$$\alpha(\mu_{P_1}, \mu_{P_2}) = \min(\mu_{P_1}, h(\mu_{P_1}, \mu_{P_2})) \qquad (2.23)$$

is sought.

From the aforementioned axioms and structure, the following operator is created [4]:

$$\alpha(\mu_{P_1}, \mu_{P_2}) = \min(\mu_{P_1}, k \cdot \mu_1 + (1-k)\mu_{P_2}), \qquad (2.24)$$

where μ_1 is the satisfaction degree to a constraint, μ_2 is the satisfaction degree to a wish and $k \in [0, 1]$ expresses relation between constraint and wish. If $k = 0$, the result is ordinal minimum t-norm; if $k = 1$, the result depends only on constraint P_1. If $k = 0.5$ is chosen, the operator is:

$$\alpha(\mu_{P_1}, \mu_{P_2}) = \min(\mu_{P_1}, \frac{\mu_{P_1} + \mu_{P_2}}{2}) \qquad (2.25)$$

At the first glance, when we consider P_1 as N and P_2 as P, then it could be the bipolar query as expressed in Sect. 2.4.4.1. However, this *and if possible* operator cannot handle bipolarity, because α does not keep both P_1 and P_2 separated. Anyway, this operator can efficiently solve many practical tasks where connective merges constraints and wishes.

Example 2.9 A client of real estate agency searches for a non expensive flat and if possible near the lake. In order to focus on *and if possible* operator, steps of constructing fuzzy sets and retrieving tuples from the database are skipped. Flats and respective membership degrees are in Table 2.9. This table illustrates satisfaction of aforementioned axioms. In case of symmetric conjunction flats $Ft6$ and $Ft7$ are indistinguishable. By operator (2.24) high satisfaction of wish is not as important as high satisfaction of constraint. Furthermore, if wish is fully non-satisfied, then the matching degree to constraint is lowered; if constraint is fully non-satisfied, then the tuple's matching degree is 0. Hence, averaging operators cannot be applied. Furthermore, when minimum function is applied both attributes are constraints.

The formula (2.24) corresponds to the local interpretation of constraints and wishes, that is, each tuple is examined independently. The first row in Table 2.8 and third row in Table 2.9 have the same membership degrees to constraints and wishes, but the calculated matching degree is higher by (2.22) than by (2.25), because in Table 2.8 this tuple dominates other tuples which is reflected in the matching degree.

Clearly, result by (2.24) for $k \neq 0$ and $k \neq 1$ does not correspond with the minimum operator and using only constraint, respectively (Table 2.9). Matching degree is approaching to the result obtained by minimum t-norm (1.47), when parameter k is approaching to the value of 0.

If we try to model preferences and wishes as conditions with different priorities (Sect. 2.4.2), then the solution is not the same. Although, some tuples could have the

Table 2.9 Non expensive flats and if possible near to the lake

Flat	$\mu_{LP}(r)$	$\mu_{SD}(r)$	$\alpha(r)(2.25)$	Min (μ_{LP}, μ_{SD})
Ft 1	1	0.7	0.85	0.7
Ft 2	0.8	0.8	0.8	0.8
Ft 3	0.8	0.5	0.65	0.5
Ft 4	1	0	0.5	0
Ft 5	0.8	0.1	0.45	0.1
Ft 6	0.6	0.1	0.35	0.1
Ft 7	0.1	0.6	0.1	0.1
Ft 8	0	1	0	0
Ft 9	0.9	0.4	0.65	0.4

where $\mu_{LP}(r)$ stands for membership degree to low price and $\mu_{SD}(r)$ to short distance

same matching degree calculated by (2.12) when weights of 1 and 0.5, correspondingly to constraint and wish in (2.25) are used (e.g. for matching degrees of 0.8 and 0.2), care should be taken, as the meanings of the queries is different. It is illustrated on tuples $Ft9$ in Table 2.9 and tuple $r1$ in Table 2.4. Even though these tuples have the same satisfaction degrees, the matching degree is different.   □

Analogously, the *or else* is dually defined, i.e. operator should be more drastic than the *or*, because P_2 is not a full alternative to P_1; less restrictive than using only P_1; must have asymmetric behaviour (the permutation of P_1 and P_2 gives different result); monotonic in both constraints and wishes; P_1 *orelse* P_2 is equivalent to P_1 *orelse* $(P_1 \text{ or } P_2)$. From these axioms the following operator appears [4]:

$$\beta(\mu_{P_1}, \mu_{P_2}) = \max(\mu_{P_1}, k \cdot \mu_1 + (1-k)\mu_{P_2}), \tag{2.26}$$

where variables have the same meaning as in (2.24). If $k = 0$, the result is ordinal maximum s-norm (1.59). If $k = 1$, the result depends only on alternative P_1. If $k = 0.5$ is chosen, the *or else* operator yields:

$$\beta(\mu_{P_1}, \mu_{P_2}) = \max(\mu_{P_1}, \frac{\mu_{P_1} + \mu_{P_2}}{2}) \tag{2.27}$$

Construction of fuzzy sets for constraints and wishes could be modelled as independent. Construction of fuzzy sets for N does not have influence on construction of fuzzy sets for P (in Example 2.9 low price for constraint and short distance for wish). It depends on users, which way for construction of fuzzy sets is chosen.

An interface covering constraints and wishes and construction of fuzzy set from data is demonstrated in Appendix A.2.

Constraints P and wishes N could contain atomic or compound conditions. A compound condition could be of any form mentioned above, e.g. *(unemployment is high and pollution is small) and if possible (population density is high and altitude is small).* As braces suggest, firstly, membership degrees inside constraint and wish are calculated. Secondly, asymmetric conjunction is applied.

2.4.5 Quantified Queries

This class of database queries uses linguistic quantifiers in query conditions. These conditions can be used as nested subqueries, especially for the relationship 1:N such as REGION-DISTRICT or CUSTOMER-ORDER. An example of such a query is *select districts where about half of municipalities have high altitude.*

The second application is in query relaxation tasks. A query usually consists of several atomic predicates merged by the *and* connective ($\wedge_{i=1}^{n} P_i$). The answer is empty, even if only one atomic predicate is not satisfied (whereas values of attribute for several tuples almost "touch" the space delimited by the predicate), or none of predicates is satisfied. When all atomic predicates must be satisfied, then the *and* connective is option. Otherwise, quantified query is a solution.

Users may be interested in tuples which meet majority of atomic predicates. In this way the query is of structure *select tuples where Q of* $\{P_1, P_2,..., P_n\}$ *is satisfied* [29] where P_i ($i = 1, ..., n$) can be either crisp or fuzzy condition and Q is a linguistic quantifier *most of*, but other quantifiers such as *about half* and *few* can be also applied.

Furthermore, constraints and wishes can be applied on quantified queries. This query is of structure Q^C *of* $\{P_i\}$ *and if possible* Q^W *of* $\{P_j\}$, where Q^C stands for quantifier appearing in constraint part of the query and Q^W stands for quantifier explaining wish part of the query. These queries can be solved by bipolar approaches or asymmetric conjunction. The former is suggested in [26] to be applied (2.22) on quantified preferences and wishes. The latter is suggested in [19] adjusting (2.24) to quantified preferences and wishes.

In addition, quantified queries mitigate empty answer problem, because these queries are less restrictive than queries created by the *and* connective. Empty answer problems are discussed in Sect. 2.5.

Example 2.10 The task is to find suitable village for building house. The relevant predicates could be: altitude above sea level around 1500 m (P_1), small population density (P_2), medium area of village size (P_3), low pollution (P_4), high number of sunny days (P_5), short distance to the region capital (P_6) and positive reviews about village (P_7). It is highly presumable that none of villages meets all predicates in a query of the structure $\wedge_{i=1}^{7} P_i$, even though predicates have flexible boundaries.

In order to solve this problem user may say that village should be considered, if it meets majority of predicates. Let us further say that P_1, P_2, P_3 and P_4 are constraints and P_5, P_6 and P_7 are wishes.

This example is solved in Chap. 3, where other parts required for coping with such a query are introduced.   □

Quantified queries expressed by linguistic quantifiers and fuzzy or crisp predicates are basic building blocks for linguistic summaries. In this case the answer is not a set of tuples and their respective matching degrees, but the truth degree of the sentence. For instance, marketing department is curious to know whether *most of middle-aged customers have short payment delay*. Because linguistic summaries are separated topic in this book, Chap. 3 is dedicated to them.

2.4.6 Querying Changes of Attributes over Time

Particular interest is focused on analysing development of attributes over time to reveal trends, changes and to forecast future developments. Theory of time series is a mature science field capable to cope with broad variety of tasks and trends. The intent of this section is not to contribute to this field, but to show how fuzzy queries can be used for retrieving tuples of preferred or critical trends. An example of such a query is *select municipalities which have high positive change (increase) in length of roads*. Similarly, the aim of query *select customers which have almost no change in amounts of orders* is to identify customers not influenced by the marketing campaign.

In these cases, queries are not focused on attributes and their respective values, but on difference between values in target year Y_t and reference (initial) year Y_i. Hence, the difference is theoretically value form the interval $[-\infty, \infty]$ with neutral element 0 (no change).

The usual terminology of expressing changes in fuzzy control is used in this section. The inspiration for construction of fuzzy sets has arisen from fuzzy controls of technical systems. Changes in the left half of the interval are labelled as negative, e.g. negative small, negative medium and negative high. Correspondingly, changes in the right part of the interval are labelled as positive small, positive medium and positive high. Finally, changes around value of 0 are labelled as almost zero. Linguistic variable *change* and possible definitions of its terms are plotted in Fig. 2.4.

In many fields, including municipal statistics or customer relationship management this way of naming could be weird. Lets us consider changes in pollution. If pollution significantly decreases, e.g. by value of $-75\,\%$, it can be hardly expressed as negative high change for inhabitants. On the other side, employment is an attribute for which the meaning of positive and negative changes are opposite. In this book we use terms as shown in Fig. 2.4 naming negative change for values lower than 0.

If changes are not stored in the database, then change is calculated as compound attribute for a tuple r in the following way:

$$A_{cr} = \frac{A_{Y_{tr}} - A_{Y_{ir}}}{A_{Y_{ir}}} \cdot 100, \tag{2.28}$$

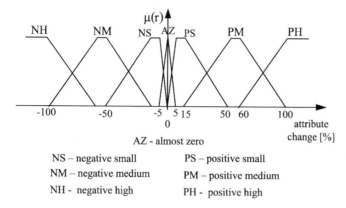

AZ - almost zero

NS – negative small   PS – positive small
NM – negative medium   PM – positive medium
NH - negative high   PH - positive high

Fig. 2.4 Fuzzification of the linguistic variable *change*

where A_{cr} is a change of tuple r for attribute A, $A_{Y_{tr}}$ is a value of attribute A in a target year and $A_{Y_{ir}}$ is a value of attribute A in an initial or referenced year. An example of this kind of query is demonstrated in Appendix A.3.

Therefore, this is an additional computational effort, but compensated with an efficient way for identifying tuples with critical or preferred changes.

Moreover, this kind of queries can be extended with all aforementioned approaches: commutative query conditions (e.g. *select municipalities which have high positive change in length of roads and high positive change in employment*), preferences between atomic conditions, constraints and wishes, quantified queries (e.g. *select regions where most of municipalities has almost no change in gas consumption*).

2.5 Empty and Overabundant Answers

Queries (either crisp or fuzzy) contain a logical condition which delimits tuples we are looking for. As a result of the query two extreme situations may occur: no data or very large amount of data satisfy the query condition. The user is confronted with informativeness of the result. In some cases, these answers are informative enough, e.g. empty answer of the condition *customers in payment delay* means that all of them meet the payment deadline and therefore no reminder should be generated.

In other cases, the main objective is to solve the problem; that is, to obtain a non-empty result to inform users, why an empty answer occurred and how close are tuples to meet the query condition. The opposite holds for an overabundant answer. A survey [7] has provided a detailed insight into these two problems.

2.5.1 Empty Answer Problem

The empty answer problem simply means that no data match the query condition. The fuzzy query Q_f (2.6) results in an empty set if $A_{Q_f} = \emptyset$. It could be useful to provide some alternative data which almost match query condition. This problem was initially recognized in [1], where linguistic modifiers were suggested as a solution. The goal of these modifiers is to slightly modify fuzzy sets in order to obtain a less restrictive variant of query condition which may return some data and remains semantically close to the initial query.

This problem appears in crisp queries more frequently due to sharp conditions (Examples 2.2 and 2.11). Making conditions flexible mitigates, but not fully eliminates this problem.

Generally, the query Q (either crisp or fuzzy) is transformed into a less restrictive variant Q_T by the transformation T of the condition. The querying process is then repeated until the answer is not empty, or the modified query condition is semantically far from the original one.

In bipolar queries and asymmetric conjunction (*and if possible*) answer is empty, if no tuple meets the constraint part. The satisfaction of wish is not so relevant for the occurrence of empty answer. Hence, transformation should be primarily focused on the constraint part.

Example 2.11 A simple example is selecting high players for a basketball team. The condition is *where height > 200 cm*. If query ends as empty, there are two possibilities: heights of all players are far from 200 cm or some of them almost meet the condition. Hence, query is transformed into a less restrictive one by transformation $T = 200-5$, i.e. relaxed query is: *where height > 195 cm*. If relaxed query returns no players, next transformation creates condition *where height > 190 cm*, and so on. The process of successive transformations should stop when some players are selected or the transformed query is far from the initial one, e.g. condition *where height > 175 cm* could be hardly considered as a selection condition for high basketball players.

The main drawback is in the satisfaction degree which is always of the value of 1 for all selected records. It is why there is no difference in membership degrees of records selected in any of transformed queries. The same problem was discussed in Sect. 2.1 in Fig. 2.1.   □

In [8] the following approaches to defeat the empty answer problem are recognized: the linguistic modifier-based approach, the fuzzy relative closeness-based approach and the absolute proximity-based approach. In these approaches the query weakening process should meet the following constraints for each predicate involved in the weakening process [7]:

$$C1: \quad \forall x \in D, \quad \mu_{T(P)}(x) \geq \mu_P(x), \tag{2.29}$$

where x is the value of the attribute A in the domain D. The transformation T does not decrease the membership degree for any tuple in domain D of attribute A for predicate P. Tuples in transformed query should have higher membership degree because, this query is less restrictive.

$$C2:\ support\ (P) = \{x \mid \mu_P(x) > 0\} \subset support\ (T(P)) = \{x \mid \mu_{T(P)}(x) > 0\}$$
(2.30)

The transformation extends the support (1.9) of fuzzy set included in the condition. In the case of trapezoidal fuzzy set in Fig. 1.7, the support is the $[a, b]$ interval constructed for the predicate P. It means that the condition is relaxed (parameter a is transformed to lower value, whereas parameter b is transformed to higher value) to retrieve more tuples.

$$C3:\ core\ (P) = \{x \mid \mu_P(x) = 1\} = core\ (T(P)) = \{x \mid \mu_{T(P)}(x) = 1\}$$
(2.31)

The transformation preserves the core (1.10) of fuzzy set. The transformed query cannot retrieve more tuples, which fully meet the relaxed condition than the initial query condition. This is a significant difference between relaxing crisp and fuzzy query. This requirement emphasizes the benefit for relaxing fuzzy queries in comparison to the relaxation of the crisp ones (Example 2.11).

An example of the original predicate (P) and the transformed one ($T(P)$) according to the requirements C1–C3 (2.29–2.31) is plotted in Fig. 2.5 for the condition expressed by the triangular fuzzy set and in Fig. 2.6 for the condition expressed by the L fuzzy set. The transformation for the trapezoidal fuzzy set is shown in [7]. If a singleton fuzzy set (Fig. 1.10) is used, then it could be transformed into less restrictive triangular fuzzy set. Requirement C3 causes that the membership degree is equal to 1 only for tuples having value of attribute equal to m (1.24). In this way, a sharp number is transformed into less restrictive triangular fuzzy number.

In order to ensure that the retrieved records are not semantically far from the initial query condition, the stopping criterion should be incorporated. In cases of the first and the third approaches aforementioned no intrinsic semantic limit is provided. Hence, the user has to specify a set of non-adequate data. In case of searching high basketball players the set for stopping criterion might be explained by the sharp limit *height not smaller than 185 cm*, i.e. $C_c = \{x \in D \mid x < 185\}$. Set C_c is defined by its

Fig. 2.5 Weakening of the condition defined by a triangular fuzzy set

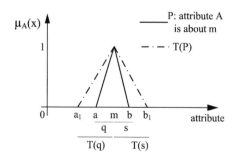

Fig. 2.6 Weakening of the condition defined by a L fuzzy set

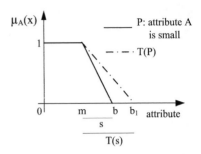

Fig. 2.7 Requirement C4 is activated for the $T(T(r))$

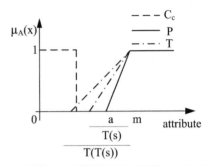

characteristic function $\varphi_c(x)$ on the domain D. Applied on the fuzzy condition, the additional requirement C4 is as follows (Fig. 2.7):

$$C4: \quad \min(\mu_{T(P)}(x), 1 - \varphi_c(x)) = 0 \qquad (2.32)$$

Concerning the requirements C1–C3 (2.29–2.31), the stopping criterion says that the weakening process stops, when the answer to modified Q_f is not empty. The constraint C4 (2.32) ensures that only tuples which are not semantically far from the initial query are selected. When during the weakening process the transformation $T(T(P)...)$ enters into the core of C_c, it causes that C4 gets value of 0 (because $1 - \varphi_c(x) = 0$) and therefore weakening process stops. In this way the relaxation is controlled.

Instead of the crisp set C_c we can construct fuzzy set F_c, a set of more or less non-adequate data.

Aforementioned approaches deal with the local weakening, that is, the basic weakening transformation applies to each atomic condition.

The further approach is focused on replacing the query which caused an empty answer by a semantically similar one which has been already processed and provided non-empty answer [2].

2.5.2 Overabundant Answer Problem

The opposite situation arises when a large amount of tuples satisfies the query condition. The query Q_f results in overabundant answer if the cardinality of A_{Q_f} (2.6) is too large.

To recognize the empty answer problem is a trivial task. In the case of an overabundant answer problem, it is the opposite. It is not easy to answer the question, where lies the boundary between non-overabundant and overabundant answers. It depends on the user, how many tuples he wants to obtain, and on the number of tuples in the database.

From the theoretical point of view the problems concerning empty and overabundant or plethoric answers are dual. The ways how to solve overabundant answer problems are examined in [7]. Generally, two situations might arise: too many tuples fully meet a query condition and/or too many tuples partially meet a query condition.

The first situation requires intensification of the query condition; that is, reduction of the core of fuzzy sets in order to reduce the number of tuples that fully meet the query condition. This process is dual to the weakening process expressed by requirements (2.29)–(2.32). At the first glance it is evident that this transformation is not applicable on triangular and singleton fuzzy sets.

Example 2.12 Case 1: Large number of customers fully meet the condition *turnover = 950*. We cannot reduce the core, because it is expressed as singleton. The possible solution is adding an additional suitable atomic condition semantically close to the initial query.

Case 2: Large number of customers fully meet the condition *turnover about 950*, where the term is expressed as a trapezoidal fuzzy set. Hence, the answer can be reduced by the intensification of the flat segment in the initial query condition.   □

At the first glance the second situation is easy to solve, because the theory of fuzzy sets offers several options. The first one is expressing the *and* operator with the Łukasiewicz t-norm or nilpotent minimum t-norm, because tuples which do not significantly match the query condition are excluded. The second option is the α-cut (1.15). For example, the *threshold* clause [6, 47] could solve this problem. However, a situation when very large number of answers has the same maximal score (different from value of 1) might occur [20]. In this case the intensification of condition is not the solution. These records will have again the same, although lower membership degree to the answer. α-cuts and the more restrictive t-norms are not applicable as well. The solution could consist in adding additional elementary condition semantically close to the initial one. An approach for adding a semantically similar attribute to the query condition in order to obtain more restrictive solution is suggested in [10]. Additional possible option are fuzzy functional dependencies [41, 46] which are useful in the process of mining related attributes. A related attribute could be connected by the *and* operator to intensify the initial query condition.

Aforementioned approaches cope with these problems mainly, when they appear: after the query realization. Another possibility is revealing information about possible risk of empty and overabundant answers. The first approach is the pre-computation of the summary of database in order to retrieve information about distribution of data in domains or parts of domains limited by the created query condition, which is about to be realized. In this way a single scan of databases provides relevant information about fuzzy cardinality [40]. The second approach is construction of fuzzy sets for linguistic conditions directly from data discussed in Sect. 2.2.

Anyway, from the practical point of view, the empty answer problem is more relevant for users than the overabundant one. The empty sheet of data provides limited information about data in databases, whereas the sheet containing a large number of tuples could be filtered by several other methods in spreadsheet software tools, for example.

2.6   Some Issues Related to Practical Realization

From the user's point of view, fuzzy queries are seen as akin to human-oriented languages, as they offer for the user to linguistically formulate query conditions, i.e. manage vagueness in queries for searching relevant tuples.

The purpose of an interface is to offer for non-expert users the ability to ask for data without necessity to know the complexity of relational database and fuzzy sets and fuzzy logic theory. There are many articles dealing with the user-friendly interfaces at disposal, e.g. [21, 27, 37, 39, 43]. The appropriate user's interface is immanent for the acceptance of flexible queries by variety of users. It means that topics discussed in this chapter need to be offered for users and managed in an efficient and appropriate way. In addition, presenting retrieved records and their matching degrees in useful and understandable ways is also a very important task. In Appendix A interfaces for managing several kinds of fuzzy queries for the municipal statistics database are discussed including presentation of solutions in tables and speculation on merging of matching degrees with polynomial areas in thematic maps.

The architecture used in this book is shown in Fig. 2.8. The mathematical background is based on the GLC (2.9), (2.10). The kernel of the application is used in next chapters.

The operations of two-valued logic meet all axioms of Boolean algebra, namely excluded middle, contradiction and idempotency, whereas operations of fuzzy logic do not meet all of them [35]. At the first glance violation of the excluded middle and contradiction axioms could be considered as problematic, because user could expect that the *and* connective between proposition and its negation should be always 0. However, uncertainty in belonging to a proposition is reflected in belonging to its negation and therefore higher value than 0 is acceptable. Although this fact is disputable from the theoretical point of view (some solutions exist how to solve it, e.g. interpolative realization of Boolean algebra [35, 36]), this is not a significant problem in querying.

The fact which is more problematic, is the lack of satisfying the idempotency axiom (1.53) for all t-norms, except the minimum t-norm (1.47). Let us look at

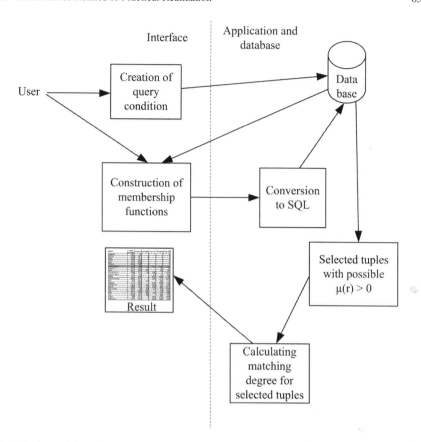

Fig. 2.8 A possible architecture of the flexible querying

two queries: *A is High* and *A is High and A is High*, where $\mu(r) < 1$. Some t-norms provide different answers to the first and the second query, namely the product (1.48) and Łukasiewicz (1.49) t-norms. Moreover, in the case when $\mu(r) < 0.5$ and using Łukasiewicz t-norm, the result is 0. This kind of query is not a realistic one, but if the user has the freedom to create fuzzy queries over a list of database attributes, these situations might lead (for example in the testing of a software application) to speculation of general applicability of fuzzy queries in the practice. This kind of query could appear either as a mistake during the construction of the condition or by purpose during testing application. There is a simple solution: check, before query realization, whether the user doubled the same atomic condition (semantics of query). If it is true, then the doubled atomic condition should be excluded from query. It especially holds, when other than minimum or nilpotent minimum t-norms are used. Developers of classical queries do not need to focus their attention on this problem, because classical conjunction is idempotent.

A powerful scenario for using flexible queries consists of the topics mentioned in Sects. 2.2, 2.4 and 2.5. In the first step, parameters of linguistic terms, preferences, bipolarities between elementary conditions, threshold and appropriate aggregation functions are created in cooperation between user and application. In this way occurrence of empty and overabundant answers is significantly mitigated. Even though, empty and overabundant answers might appear. In this case, software and user should cooperate to solve these problems.

References

1. Andreasen, T., Pivert, O.: On the weakening of fuzzy relational queries. In: 8th International Symposium on Methodologies for Intelligent Systems, pp. 144–153, Charlotte (1994)
2. Bosc, P., Brando, C., Hadjali, A., Jaudoin, H., Pivert, O.: Semantic proximity between queries and the empty answer problem. In: Joint IFSA-EUSFLAT Conference, pp. 259–264, Lisbon (2009)
3. Bosc, P., Pivert, O.: On a fuzzy bipolar relational algebra. Inf. Sci. **219**, 1–16 (2013)
4. Bosc, P., Pivert, O.: On four noncommutative fuzzy connectives and their axiomatization. Fuzzy Sets Syst. **202**, 42–60 (2012)
5. Bosc, P., Pivert, O.: SQLf query functionality on top of a regular relational database management system. In: Pons, O., Vila, M.A., Kacprzyk, J. (eds.) Knowledge Management in Fuzzy Databases, pp. 171–190. Springer, Berlin, Heidelberg (2000)
6. Bosc, P., Pivert, O.: SQLf: a relational database language for fuzzy querying. IEEE Trans. Fuzzy Syst. **3**, 1–17 (1995)
7. Bosc, P., Hadjali, A., Pivert, O.: Empty versus overabundant answers to flexible relational queries. Fuzzy Sets Syst. **159**, 1450–1467 (2008)
8. Bosc, P., Hadjali, A., Pivert, O.: Weakening of fuzzy relational queries: and absolute proximity relation-based approach. Mathware Soft Comput. **14**, 35–55 (2007)
9. Bosc, P., Pivert, O., Smits, G.: On a fuzzy group-by and its use for fuzzy association rule mining. In: 14th East-European Conference on Advances in Databases and Information Systems (ADBIS'10), pp. 88–102, Novi Sad (2010)
10. Bosc, P., Hadjali, A., Pivert, O., Smits, G.: An approach based on predicate correlation to the reduction of plethoric answer sets. In: Guillet, F., Ritschard, G., Zighed, D.A. (eds.) Advances in Knowledge Discovery and Management, Studies in Computational Intelligence, vol. 398, pp. 213–233. Springer, Berlin, Heidelberg (2012)
11. Chamberlin, D.D., Boyce, R.F.: SEQUEL: A structured english query language. In: The ACM SIGMOD Workshop on Data Description, Access and Control, pp. 249–264, Ann Arbour (1974)
12. Cox, E.: Fuzzy Modeling and Genetic Algorithms for Data Mining and Exploration. Morgan Kaufman, San Francisco (2005)
13. Date, C.J., Darwen, H.: A Guide to SQL Standard, 4th edn. Addison-Wesley, Boston (1996)
14. Dubois, D., Prade, H.: Handling bipolar queries in fuzzy information processing. In: Galindo, J. (ed.) Handbook of Research on Fuzzy Information Processing in Databases, pp. 97–114. Information Science Reference, Hershey (2008)
15. Dubois, D., Prade, H.: Using fuzzy sets in flexible querying: why and how? In: Andreasen, T., Christiansen, H., Larsen, H.L. (eds.) Flexible Query Answering Systems, pp. 45–60. Kluwer Academic Publishers, Dordrecht (1997)
16. Dubois, D., Prade, H.: Weighted minimum and maximum operations in fuzzy set theory. Inf. Sci. **39**, 205–210 (1986)
17. Garibaldi, J.M., John, R.I.: Choosing membership functions of linguistic terms. In: 12th IEEE International Conference on Fuzzy Systems (FUZZ'03), pp. 578–583, St. Louis (2003)

18. Gupta, M.M., Oi, J.: Theory of t-norms and fuzzy inference methods. Fuzzy Sets Syst. **40**, 431–450 (1991)
19. Hudec, M.: Constraints and wishes in quantified queries merged by asymmetric conjuction. In: International Conference Fuzzy Management Methods (ICFM square 2016), Fribourg (2016)
20. Hudec, M., Vučetić, M.: Some issues of fuzzy querying in relational databases. Kybernetika **51**, 994–1022 (2015)
21. Hudec, M.: Improvement of data collection and dissemination by fuzzy logic. In: Joint UNECE/Eurostat/OECD Meeting on the Management of Statistical Information Systems (MSIS 2013), Paris-Bangkok (2013)
22. Hudec, M.: Dynamically modelling of fuzzy sets for flexible data retrieval. In: 46th Scientific meeting of the Italian statistical society, Rome (2012)
23. Hudec, M.: Fuzzy improvement of the SQL. Yugoslav. J. Oper. Res. **21**, 239–251 (2011)
24. Hudec, M.: An approach to fuzzy database querying, analysis and realisation. Comput. Sci. Inf. Syst. **6**, 127–140 (2009)
25. Hudec, M., Sudzina, F.: Construction of fuzzy sets and applying aggregation operators for fuzzy queries. In: 14th International Conference on Enterprise Information Systems (ICEIS 2012), Proceedings, vol. 1, pp. 253–257, Wroclaw (2012)
26. Kacprzyk, J., Zadrożny, S.: Compound bipolar queries: combining bipolar queries and queries with fuzzy linguistic quantifiers. In: 8th Conference of the European Society of Fuzzy Logic and Technology (EUSFLAT 2013), pp. 848–855, Milan (2013)
27. Kacprzyk, J., Zadrożny, S.: Computing with words in intelligent database querying: standalone and internet-based applications. Inform. Sci. **134**, 71–109 (2001)
28. Kacprzyk, J., Zadrożny, S.: FQUERY for Access: fuzzy querying for windows-based DBMS. In: Bosc, P., Kacprzyk, J. (eds.) Fuzziness in Database Management Systems, pp. 415–433. Physica-Verlag, Heidelberg (1995)
29. Kacprzyk, J. Ziółkowski, A.: Database queries with fuzzy linguistic quantifiers. IEEE Trans. Syst. Man Cyber. **SMC-16**, 474–479 (1986)
30. Kacprzyk, J., Pasi, G., Vojtáš, P., Zadrożny, S.: Fuzzy querying: issues and perspectives. Kybernetika **36**, 605–616 (2000)
31. Klement, E.P., Mesiar, R., Pap, E.: Triangular norms. Position paper I: basic analytical and algebraic properties. Fuzzy Sets Syst. **143**, 5–26 (2004)
32. Klir, G., Yuan, B.: Fuzzy Sets and Fuzzy Logic. Theory and Applications. Prentice Hall, New Jersey (1995)
33. Lacroix, M., Lavency, P.: Preferences: putting more knowledge into queries. In: 13th International Conference on Very Large Databases, pp. 217–225, Brighton (1987)
34. Pivert, O., Bosc, P.: Fuzzy Preference Queries to Relational Databases. Imperial College Press, London (2012)
35. Radojević, D.: Interpolative realization of Boolean algebra as a consistent frame for gradation and/or fuzziness. In: Nikravesh, M., Kacprzyk, J., Zadeh, L.A. (eds.) Forging new frontiers: fuzzy pioneers II, Studies in fuzziness and soft computing, pp. 295–317. Springer, Berlin, Heidelberg (2008)
36. Radojević, D.: [0, 1] Valued logic: a natural generalization of Boolean logic. Yugoslav J. Oper. Res. **10**, 185–216 (2000)
37. Ribeiro, R.A., Moreira, A.M.: Fuzzy query interface for a business database. Int. J. Human-Comput. Stud. **58**, 363–391 (2003)
38. Rosado, A., Ribeiro, R.A., Zadrożny, S., Kacprzyk, J.: Flexible query languages for relational databases: an overview. In: Bordogna, G., Psaila, G. (eds.) Flexible Databases Supporting Imprecision and Uncertainty. Studies in fuzziness and soft computing, vol. 203, pp. 3–53. Springer, Berlin, Heidelberg (2006)
39. Smits, G., Pivert, O., Girault, T.: ReqFlex: Fuzzy queries for everyone. In: 39th International Conference on Very Large Data Bases, pp. 1206–1209, Trento (2013)
40. Smits, G., Pivert, O., Hadjali, A.: Fuzzy cardinalities as a basis to cooperative answering. In: Pivert, O., Zadrożny, S. (eds.) Flexible Approaches in Data, Information and Knowledge Management. Studies in Computational Intelligence, vol. 497, pp. 261–289. Springer International Publishing Switzerland (2014)

41. Sözat, M.I., Yazici, A.: A complete axiomatization for fuzzy functional and multivalued dependencies in fuzzy database relations. Fuzzy Sets Syst. **117**, 161–181 (2001)
42. Tahani, V.: A conceptual framework for fuzzy query processing: a step toward very intelligent database systems. Inf. Process. Manag. **13**, 289–303 (1977)
43. Tudorie, C.: Intelligent interfaces for database fuzzy querying. Ann Dunarea de Jos Univ. Galati, Fascicle III, **32**(2), Galati (2009)
44. Tudorie, C.: Qualifying objects in classical relational database querying. In: Galindo, J. (ed.) Handbook of Research on Fuzzy Information Processing in Databases, pp. 218–245. Information Science Reference, Hershey (2008)
45. Urrutia, A., Pavesi, L.: Extending the capabilities of database queries using fuzzy logic. In: Collecter-LatAm conference, Santiago (2004)
46. Vučetić, M., Vujošević, M.: A literature overview of functional dependencies in fuzzy relational database models. Tech. Technol. Educat. Manag. **7**, 1593–1604 (2012)
47. Wang, T-C., Lee, H-D., Chen, C-M.: Intelligent queries based on fuzzy set theory and SQL. In: 39th Joint Conference on Information Science, pp. 1426–1432, Salt Lake City (2007)
48. Zadrożny, S., Kacprzyk, J.: Bipolar queries: a way to enhance the flexibility of database queries. In: Ras, Z.W., Dardzinska, A. (eds.) Advances in Data Management, Studies in Computational Intelligence, vol. 223, pp. 49–66. Springer, Berlin, Heidelberg (2009)
49. Zadrożny, S., de Tré, G., de Caluwe, R., Kacprzyk, J.: An overview of fuzzy approaches to flexible database querying. In: Galindo, J. (ed.) Handbook of Research on Fuzzy Information Processing in Databases, pp. 34–55. Information Science Reference, Hershey (2008)

Chapter 3
Linguistic Summaries

Abstract In many tasks, users are not interested in data stored in relational databases, but in summarized relational knowledge and "abstracts" from the data which are expressed in a useful and understandable way by linguistic terms. Linguistic Summaries (LSs) are able to express the knowledge in the data that is concise and easily understandable by users. LSs are quantified sentences of natural language such as *most of municipalities of high altitude and low pollution have small number of respiratory diseases*. The truth value of summaries gets values from the unit interval as it is common in the fuzzy logic world. We start with simple LSs and continue with more complex ones. In this direction, selecting appropriate t-norms for aggregation and quality measures are discussed. Furthermore, a system for calculating summaries will not work properly, if it uses ill-defined membership functions. Focus is also on constructing these functions for summarizers, restrictions and quantifiers. The quality measures are also analysed, because the high truth value of sentence is not always a sufficient measure. Finally, possible applications are considered.

3.1 Benefits and Protoforms of Linguistic Summarization

Retrieving tuples from databases is the topic of Chap. 2. In many other tasks users are not interested in data, but in summaries which briefly explain data and relations among attributes.

Summarization can be realized by statistical methods. These methods summarize the essential information from a data set into few numbers [9]. Methods such as means, medians and deviations provide valuable information, e.g. *in 2011 municipalities produced waste of the average amount of 217.9184 kg per inhabitant with standard deviation of 189.2839*. However, interpreting data in this way is practicable for rather small specialized groups of people. When the quality of collected data is not high (e.g. errors in data collection or rough estimation), the calculations should fight with these issues. Hence, the following quotation holds:

> ... method of summarization would be especially practicable if it could provide us with summaries that are not as terse as the mean [48].

© Springer International Publishing Switzerland 2016
M. Hudec, *Fuzziness in Information Systems*,
DOI 10.1007/978-3-319-42518-4_3

Expressing data summarization by two-valued logic is limited. The truth value of sentence (predicate) created by the universal quantifier ($\forall$) is 1, only if all tuples meet the requirement (condition), e.g. *all territorial units have length of local roads < 200 km*. If the truth value is 0, then we do not know whether 1 or 99 % of tuples do not meet the requirement. The same comment about data quality in statistical methods holds here. If someone, who is responsible for the data collection, value of 198.92 km rounds to 200 km and moreover, no other territorial unit has length of road greater or equal 200 km, then the truth value of this sentence is 0.

Keeping aforementioned facts in mind, data summarization by fuzzy logic could be a suitable option. An example of such a summary is *most of middle-aged customers have short payment delay*. The sentences of such a structure were introduced in [47]. Because summarized information from the data is sought, these sentences are called Linguistic Summaries (LSs). This field is under deep interest of scientists and practitioners due to large variety of possible theoretical improvements and practical applications, e.g. [6, 7, 17, 21, 23, 28, 37, 39–41, 43, 50].

Graphical interpretation is also a valuable way of data summarization but cannot be always effectively applied [31]. Linguistics is an interesting alternative, when data is hard to show graphically [49]. Furthermore, linguistically summarized sentences can be read out by a text-to-speech synthesis system. This way is especially suitable, when the visual attention should not be disturbed [3].

LSs are quantified sentences of natural language that distil the most relevant information from large number of tuples and present it in humanly consistent forms. Nowadays, LSs become more important due to the exponentially growing amounts of collected data and due to the issues related to efficiently grasping these data. Individual's natural capability to grasp all necessarily information required for managing and control variety of tasks is limited what means that the need for computational support is obvious [37]. Kacprzyk and Zadrożny [23] emphasized that

> Data summarization is one of basic capabilities needed by any 'intelligent' system.

Data summarization by fuzzy logic simulates the human capability to make conclusions without precise measurements and calculations. Journalist can relatively easily find, which colour dominates on tall individuals at a party to explain the fashion trend for tall persons, for example. Different hues of colours as well as the meaning of vague term *tall persons* are not limitations for solving this task. However, if we wish to know, which of the following two sentences: *most of young commuters commute short distances; most of middle-aged commuters commute short distances* better explains the commuting behaviour, then we have to adapt these questions to database query languages. Furthermore, summarization should treat numeric as well as nonnumeric data, which is also possible with LSs as is shown later.

Linguistically quantified sentences are written in general form

$$Qx(P(x)) \tag{3.1}$$

where Qx is a linguistic quantifier, $P(x)$ is a predicate depicting evaluated attributes. The truth value of a summary is also called validity (v) and gets values from the unit interval by agreement.

In order to use advantages of SQL and linguistic summaries, Rasmussen and Yager [41] have created SummarySQL language. Further realizations are the extension of FQUERY [24] and SAINTETIQ [39, 45].

Prototype forms (protoforms) of linguistic summaries can be divided into three main groups [31]: classic protoforms, protoforms of time series and temporal protoforms.

Classic protoforms are useful for mining knowledge among attributes in traditional relational databases. These summaries can be further divided into basic structures of LS, which express information about particular attributes on the whole data set, i.e. *Q entities are S*, and into structures with restriction, which express relational knowledge among attributes on the part of a database delimited by the flexible (or sharp) restriction, i.e. *Q R entities are S*. The former is illustrated by the sentence such as *most of houses have high gas consumption*. An illustrative example of the latter is *most of old houses have high gas consumption*.

Protoforms of time series linguistically express behaviour of attributes over time. These summaries are divided into summaries describing one time series [26], such as *most stable trends are of low variability*, and into summaries considering several time series together [1]. The latter protoform is of the structure *Q R are S Q_T time*, such as *about half customers have small number of orders most of the time*. They are suitable for data warehouses, because in these data structures one of dimensions is time. Each attribute's value gets distinct time stamp, which is a valuable help for mining summaries.

Temporal protoforms do not use fuzzy quantifiers, but mode of behaviour creating summaries such as *regularly entities are S*. This kind of summaries is illustrated by a sentence such as *regularly customers buy small packages*. The term *regularly* describes the extent, which a summary holds to, considering a special temporal adjustment [36].

This section is focused on classic protoforms. In Chap. 5 these prototorms are extended for summarizing from fuzzy data stored in fuzzy relational databases.

3.2 The Basic Structure of LS

A basic LS is of the structure *Q entities in database are (have) S*, where Q is the relative quantifier (most of, about half, few, etc.) and S is a summarizer expressed by linguistic terms. An example of this summary is *most of municipalities have medium water consumption per inhabitant*.

The validity is computed in the following way [23]:

$$v(Qx(P(x))) = \mu_Q(\frac{1}{n}\sum_{i=1}^{n}\mu_S(x_i)) \qquad (3.2)$$

where n is the scalar cardinality of a database (number of tuples), $\frac{1}{n}\sum_{i=1}^{n}\mu_S(x_i)$ is the proportion of objects in a data set that satisfies $P(x)$ and $\mu_Q(y)$ is the membership function of a chosen relative quantifier shown in Fig. 1.20, where $y = \frac{1}{n}\sum_{i=1}^{n}\mu_S(x_i)$. Quantifiers describe extent to which summarizer holds for the considered data set.

The validity can be expressed by scalar cardinality as [37]

$$v(Qx(P(x))) = \mu_Q(\frac{\text{card}(S)}{n}) \tag{3.3}$$

where $\sum_{i=1}^{n}\mu_S(x_i)$ represents scalar cardinality of summarizer S: $\text{card}(S)$.

In case of the basic LS, the predicate $P(x)$ consists of the summarizer S only. Therefore we can use both notations $\mu_S(x_i)$ and $\mu_P(x_i)$. Summarizer could concern more than one atomic condition joined by the *and* connective [13]. If summarizer consists of several atomic predicates, $\mu_S(x_i)$ is calculated in the following way:

$$\mu_S(x_i) = t(\mu_{S_j}(x_i)) \tag{3.4}$$

where S_j is the j-th atomic predicate $j = 1, \ldots, m$ of summarizer S and t is a t-norm.

Atomic predicates in summaries may be crisp, as well as fuzzy ones such as *most of houses are old and have size > 500 m²*.

Naturally, the question of suitable t-norm appears again. We believe that for the summarization only the minimum t-norm is the suitable option, because the value of aggregation is not lower than the lowest satisfaction degree of atomic conditions (1.47). In this way, the proportion and therefore the validity is not artificially lowered. For example, when a tuple meets four atomic conditions with the value of 0.5, the tuple participates in the proportion with the expected value of 0.5. In case of product t-norm (1.48), this tuple participates with the value of 0.0625. Nilpotent minimum t-norm (1.54), which also shows its advantages in fuzzy queries, is not suitable for summaries. It is easy to prove. Let tuple meet the first atomic condition with 0.5. If this tuple meets the second atomic condition with 0.49, then it does not participate in proportion, whereas if tuple meets the second atomic condition with 0.51, then it participates in proportion with 0.5. The further discussion related to selection of suitable t-norms is in Sect. 3.6, where other aspects regarding quality of summaries are examined.

3.3 Relative Quantifiers in Summaries

The validity of LSs (3.2) depends on the membership function of the chosen relative quantifier, among others.

For a regular non-decreasing quantifier (e.g. *most of*) its membership function should meet the following properties:

$$y_1 \leq y_2 \Rightarrow \mu_Q(y_1) \leq \mu_Q(y_2); \quad \mu_Q(0) = 0; \quad \mu_Q(1) = 1 \tag{3.5}$$

Existing approaches for data summarization via linguistic quantifiers have manifested various quality indicators for quantifier selection [14]. For practical applications and keeping in mind (3.5) the quantifier *most of* might be given as [23, 50]

$$\mu_Q(y) = \begin{cases} 1, & \text{for } y \geq 0.8 \\ 2y - 0.6, & \text{for } 0.3 < y < 0.8 \\ 0, & \text{for } y \leq 0.3 \end{cases} \qquad (3.6)$$

This quantifier can be parametrized in the following way [16]:

$$\mu_Q(y) = \begin{cases} 1, & \text{for } y \geq n \\ \frac{y-m}{n-m}, & \text{for } m < y < n \\ 0, & \text{for } y \leq m \end{cases} \qquad (3.7)$$

Assigning value 0.5 to m could be suitable option, because linguistic term *most of* should cover proportions, which are significantly high. The interactive option is offering possibility for users to adjust parameters in the same ways, as is shown in Sect. 2.2. Anyway, if we evaluate all short sentences over a database by one fixed quantifier, then validities of these sentences will be correctly ranked.

The restrictive version of quantifier *most of* can be expressed as the quantifier *almost all*, when $n = 1$ and m gets, for instance, value 0.85. In the most restrictive case, i.e. $m = n = 1$, the crisp quantifier *all* is reached. These three quantifiers are shown in Fig. 3.1. Function marked as a low density dotted line is the least strict one. Function marked as a high-density dotted line represents more restrictive quantifier. Function marked as a solid line stands for the extremely strict quantifier—the crisp quantifier $\forall$. If only one record (from a very large number of records) does not meet the predicate, the truth value of this quantifier is 0. The same truth value is calculated when only few records meet the predicate. In case of fuzzy quantifiers this distinction is detectable.

Analogously, nonincreasing quantifier (e.g. *few*) could be created as a "mirror picture" of (3.5) and (3.6) [17].

Fig. 3.1 Adjusting the meaning of term *the majority of*

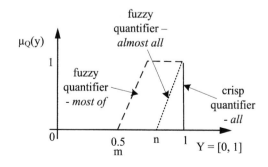

Fig. 3.2 Quantifier *about half*

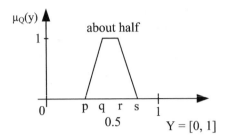

The quantifier *about half* (Fig. 3.2) should be symmetric fuzzy number around the value of 0.5, i.e. $|q - p| = |s - r|$ and $|0.5 - q| = |r - 0.5|$, when trapezoidal number is used. Triangular fuzzy number is also applicable, but it is more restrictive than the trapezoidal one, i.e. membership degree is equal to 1 only for the proportion having value of 0.5 ($q = r = 0.5$). Generally, the validation rule for this quantifier consists of additional relation, e.g. $0.25 \leq p \leq q \leq 0.5 \leq r \leq s \leq 0.75$, where values of 0.25 and 0.75 could be adjusted for particular tasks.

Example 3.1 A sport journalist wants to know, whether most of ski jumpers are tall and slim. The quantifier *most of* is expressed with parameters $m = 0.6$ and $n = 0.85$ (Fig. 3.1). Tall jumpers are expressed by the R fuzzy sets (Fig. 1.9) with parameters $a = 180$ cm and $m = 190$ cm. Slim jumpers are expressed by the L fuzzy set (Fig. 1.8) with parameters $m = 65$ kg and $b = 75$ kg. Jumpers, their height, weight and matching degree to *tall and slim* by minimum (1.47) and product (1.48) t-norm are in Table 3.1.

For the minimum t-norm the proportion of tuples satisfying condition is $P(x) = 0.695$, whereas for product t-norm it is 0.588. It implies that validity of summary is $v(Qx(P(x))) = 0.557$ for aggregating atomic conditions by minimum t-norm and 0.251 for aggregating by product t-norm. If minimum t-norm is used, the journalist would say that the statement is more or less significant (true). Otherwise, if product t-norm is used, journalist would say that the statement is insignificant. Therefore, care should be taken when working with summarizers consisted of more than one atomic condition. Furthermore, the values of proportion and validity suggest that quantifier *about half* may be better option and should be considered. □

Table 3.1 contains few tuples. The main intent was to illustrate the power of linguistically summarizing data. Obviously, the advantage of LSs is in summarizing large data sets.

Example in Appendix B illustrates possible interface for mining basic LS from the municipal database.

Table 3.1 Ski jumpers and their attributes

Jumper	Height (cm)	Weight (kg)	μ_{tall}	μ_{slim}	Minimum t-norm	Product t-norm
J1	189	66	0.9	0.9	0.9	0.81
J2	172	72.5	0	0.25	0	0
J3	194	62	1	1	1	1
J4	187.5	67.5	0.75	0.75	0.75	0.5625
J5	182.5	72.5	0.25	0.25	0.25	0.0625
J6	188	67	0.8	0.8	0.8	0.64
J7	185	70	0.5	0.5	0.5	0.25
J8	187.5	67.5	0.75	0.75	0.75	0.5625
J9	191	64	1	1	1	1
J10	193	61	1	1	1	1

3.4 LS with Restriction

A more complex type of LS is summary with restriction having the form *Q R entities in database are (have) S*, where *R* delimits part of database of interest. An example of such a summary is: *most of low polluted municipalities have low number of respiratory diseases*. The procedure for calculating truth value has the following form [41]:

$$v(Qx(P(x))) = \mu_Q\left(\frac{\sum_{i=1}^{n} t(\mu_S(x_i), \mu_R(x_i))}{\sum_{i=1}^{n} \mu_R(x_i)}\right) \tag{3.8}$$

where $\dfrac{\sum_{i=1}^{n} t(\mu_S(x_i), \mu_R(x_i))}{\sum_{i=1}^{n} \mu_R(x_i)}$ is the proportion of the records in database that meet the *S* and belong to the *R*, *t* is a t-norm, $\mu_Q(y)$ is the membership function of the chosen quantifier.

The validity can be also expressed by scalar cardinality as [37]

$$v(Qx(P(x))) = \mu_Q\left(\frac{\text{card}(S \cap R)}{\text{card}(R)}\right) \tag{3.9}$$

where $\sum_{i=1}^{n} t(\mu_S(x_i), \mu_R(x_i))$ is represented by the scalar cardinality of intersection between *S* and *R*.

Restriction *R* could be atomic or composed of several atomic conditions. The role of restriction is to focus on particular flexible part of a database. If *R* contains several atomic conditions, then the same discussion holds as for (3.4). This kind of LS is schematically depicted in Fig. 3.3. The grey areas around solid line between sets *small*, *medium* and *high* emphasizes the uncertainty areas, i.e. parts of domains, where unambiguous belonging to a particular set cannot be arranged. For example,

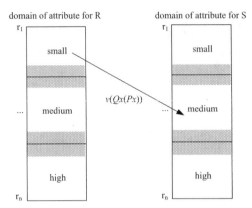

Fig. 3.3 LS with a restriction. The grey areas emphasize the uncertainty areas (overlapped fuzzy sets)

if the delimiting line between small and medium pollution is 50 mg of the measured pollutant, then 49.7 and 50.2 belong to both sets with slightly different membership degrees.

Example 3.2 The goal is to find out, whether most of expensive books have small number of pages. Books and their parameters are in Table 3.2.

Concerning the fuzzy set *high price*, the support starts on 80 € and the core begins in 90 €. Regarding the *small number of pages*, the core ends in 200 pages and support ends in 250 pages. The proportion of tuples satisfying summary (3.8) is

$$\frac{\sum_{i=1}^{n} t(\mu_S(x_i), \mu_R(x_i))}{\sum_{i=1}^{n} \mu_R(x_i)} = \frac{0.5 + 0.1 + 0.75 + 0.25 + 1}{0.8 + 0.1 + 0.75 + 0.25 + 1} = 0.895.$$

The next step is inserting this value into quantifier *most of* expressed by parameters $m = 0.5$ and $n = 0.85$. Obviously, the validity is 1, which leads to full acceptance of the tested statement. But, one should be careful in making conclusions, because other aspects may influence the solution.   □

Usually, at the beginning user does not exactly know, how to construct fuzzy sets in LSs and what can be expected from a database [32]. Care should be especially taken, when working with LSs containing restriction part. As was mentioned earlier (Sect. 2.2), the collected data are often situated only in a part of the attributes' domains. When two summaries have validity equal to 0, two possible situations might arise [18] (i) parts of domains included in LS have insufficient tuples for calculation of relations between data (or are empty); (ii) parts of domains contain tuples, but there is no significant relation between them. So, it is unclear, which interpretation is the correct one. Moreover, one should be very careful, when no tuple meets the

Table 3.2 Books, their prices and number of pages

Book	Price (€)	Number of pages
B1	88	225
B2	81	210
B3	87.5	162
B4	52	210
B5	63	295
B6	82.5	220
B7	76	230
B8	92	188
B9	72	290

R part, because it leads to dividing by zero in (3.8). This type of summary is also analysed in Sect. 3.6 focused on the quality of summaries.

3.5 Mining Linguistic Summaries of Interest

Generally, there are two main ways for extracting summaries from the data. In the first way the user creates particular linguistic summary(ies) or sentence(s) of interest for evaluating its (their) validity (Examples 3.1 and 3.2). The second way is based on automatic generation of summaries from the data.

According to [34]

> Summarization is the process of distilling the most important information from a source (or sources) to produce an abridged version for a particular user (or users) and task (or tasks).

This definition means that the goal of summarization is to briefly describe characteristics appearing in the data. In case of LSs, it can be expressed as an operations research task [33]

$$\begin{aligned} &\text{find } Q, S, R \\ &\text{subject to} \\ &Q \in \overline{Q}, R \in \overline{R}, S \in \overline{S}, v(Q, S, R) \geq \beta \end{aligned} \tag{3.10}$$

where $\overline{Q}$ is set of quantifiers of interest, $\overline{R}$ and $\overline{S}$ are sets of relevant linguistic expressions for restriction and summarizer respectively, and β is threshold value from the $(0, 1]$ interval. Each feasible solution produces a linguistic summary ($Q^* R^*$ are S^*).

If validities of all LSs are equal to 0 or are under the threshold value β, we do not cope with the empty answer problem discussed in Sect. 2.5. It just means that data are randomly distributed inside their respective domains without any relationship or grouping into the particular segments of respective domains.

The following definition also explains LSs

Linguistic summarization is a process of constructing abstract from relatively large data sets using predefined linguistic term sets and fuzzy logic.

Example 3.3 An illustrative database contains the following 7 customers with their ages {C1:26, C2:28, C3:32, C4:40, C5:54, C6:56, C7:57}. The goal is to reveal all basic LSs (3.2) by (3.10). In this case the R part is not included. The terms set of summarizers is $\overline{S}$ = {young, middle-aged, old}. The terms set expressing quantifiers $\overline{Q}$ consists of linguistic terms *few*, *about half* and *most of*. The terms sets of quantifiers and summarizers are plotted in Fig. 3.4. Parameters for all terms of the terms set $\overline{Q}$ are constructed by the uniform domain covering method (2.7), (2.8) where $L = 0$, $H = 1$ and $\theta = 2\varepsilon$. The terms set of the variable *age* is created by user. For illustrative purpose, all possible LSs and their validities are shown in Table 3.3. When the threshold value $\beta = 0.75$ (3.10) is applied, then only LSs marked as bold are shown to users. In this way LSs are able to reveal "abstract" from large data sets to support decisions.

If we further reveal that only *most of middle aged customers have high turnover and low payment delay*, then this is a relevant information for the decision: marketing department should attract more middle-aged customers. Furthermore, instead of a long list of customers and their attributes, decision makers see quantified sentences explaining the data.

In the customer relationship management (CRM) one of goals is to understand customer behaviour and therefore create marketing actions, which may improve retail. By help of LSs, marketing department has valuable information for adjusting advertisement strategies to improve profit, for example. □

Fig. 3.4 Terms sets of quantifier and summarizer

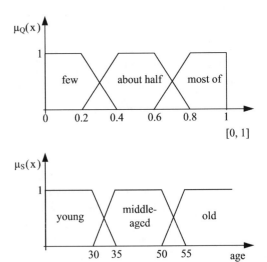

Linguistic summary	$v(Qx(Px))$
Few customers are young	0.1430
Few customers are middle aged	**0.8575**
Few customers are old	0.0000
About half customers are young	**0.8570**
About half customers are middle aged	0.1425
About half customers are old	**1.0000**
Most of customers are young	0.0000
Most of customers are middle aged	0.0000
Most of customers are old	0.0000

Table 3.3 Validities of LSs created from the predefined linguistic terms sets

The companies' data in Example 3.3 can contain further business data such as sales by particular parts of days, types of products, etc. Having all these data, we can reveal further relevant summaries. The summaries usually involve data from the companies' own databases. However, no company operates without interaction with the surrounding world. Therefore, some external data (e.g. climatic conditions, unemployment) should be taken into account, when expressing number of purchased items, for example. Hence, external data sources such as official statistics can be valuable for companies.

Since the fully automatic generation of all relevant LSs is not an easy task, there is a room for further research and development. Presumably, first results in this direction have been presented in [44] for generating LSs related to energy consumption data.

The following question naturally arises: how to efficiently calculate LSs from the large data sets? When the number of tuples and their attributes is relatively large, the computation might take much time and might be costly. For instance, when we have 2891 tuples described by 804 attributes, it is necessary to compute $2891 \cdot 804$ membership degrees according to [37]. We can avoid such an amount of computation by selecting only tuples, which at least partially belong to summarizer and restriction (Sect. 2.3 and Example B.1). When sets $\overline{Q}$, $\overline{R}$ and $\overline{S}$ contain three fuzzy sets each, then 27 possible combinations (3^3) exist. If we examine summaries only for two attributes, 27 summaries should be evaluated. When 804 attributes are considered, where R and S consists of atomic predicates, all possible variations without repetitions is $V_{n,m} = \frac{n!}{(n-m)!} = \frac{800!}{798!} = 639\ 200$. The total number of LSs is 17 258 400.

Practice and experiments show that usually not all of these summaries have high validity. Hence, the computation of summaries with low validity is pointless. Admittedly, validities of already processed summaries can be used for excluding calculation of validities still in queue. For instance, if validity of LS: *few customers are middle aged* is very high (Table 3.3), then we can hardly expect that validity of *most of customers are middle aged* is significantly higher than 0. In addition, the value of proportion helps in process of excluding irrelevant summaries. For instance, if the proportion of tuples in (3.8) is equal to 0.51, then it indicates that significant validity can be expected for the quantifier *about half*.

In this direction two aspects might help. The first is based on user's needs and attitude. The user decides, which attributes are relevant for a particular task and which combination of Q, R and S might result in significant validity. This decision is based on common sense and perceptions, e.g. validity of summary *about half of historically old municipalities have high water consumption* is irrelevant for analysing unemployment. The second aspect is based on pre-computations either by cardinalities [42] or constructing summaries for attributes which display a preselected query of nonzero level [13].

Example 3.4 Researchers are interested to find out, whether significant summarizations between the attributes *population density* and *production of waste* exist. This example is taken from [18]. In the first step, both attributes are fuzzified into three fuzzy sets (small, medium, high) by the uniform domain covering method (2.7) and (2.8). Applying the quantifier *majority of* defined with parameters $m = 0.3$ and $n = 0.8$ (a less restrictive variant of the quantifier *most of*) the validities for three LSa are calculated and shown in Table 3.4.

Maximal number of possible LSs in this example is nine. However, other six possible combinations of restrictions and summarizers may have very low validity and therefore were not evaluated.   □

This experiment can be relevant for policy or business decision-making in the area of waste management. If an experiment is focused on relation between the attributes *the year of the first written notice* and *size of agricultural land*, then it might be interesting for, e.g. historians. Therefore, we can mine relational knowledge among relevant attributes for variety of research fields, decision and policy-making.

Table 3.4 Validities of LSs created from the population density and waste production in municipalities

Linguistic summary	$v(Qx(Px))$
Majority of municipalities having small population density have small production of waste	1
Majority of municipalities having high population density have high production of waste	0.662
Majority of municipalities having medium population density have medium production of waste	0.132

3.6   Quality Measures of LSs

We could say that

linguistic summary is a more or less accurate textual description (summary, abstract) of a database satisfying certain user's requirements.

This simple definition hides many challenges some of which were discussed above. It is worth noting that summaries bear more challenges than fuzzy queries. Roughly speaking, data either meet (fully or partially), or do not meet the query criterion and therefore are selected, or not. So, when the data are retrieved, the task is more or less finished. On the contrary, obtained validities of summaries (it especially holds for LSs with restriction part) is not always the end of a task. The next step is measuring the quality of calculated summaries. LSs might be trapped into outliers, there might be better description of data even with lower value of validity, subjectivity in constructing fuzzy sets might affect validity and the like. Hence, some quality measures should be considered in order to mitigate vagueness of validity.

In addition, mining summaries is not always the main goal of a task. For instance, summaries with restriction part can be transformed into fuzzy IF-THEN rules. Thus, the quality of summaries influences the quality of created rule base.

The main objectives, when studying quality of summaries are (i) to see whether a particular LS is of sufficient quality to summarize considered database and (ii) to compare two or more LSs created on the same database. For both objectives, the common task is to define a criterion for measuring quality. Regarding the latter objective, a binary ordered relation in the space of possible LSs is required. This relation should at least meet the properties of reflexivity and transitivity [10]. The relation is more strict, if it meets the property of anti-symmetricity. A criterion usually consists of several atomic quality measures. But, different tasks and perspectives require diverse measures and even differently defined same atomic measure. In this section, two views on quality and their respective measures are examined.

3.6.1   Quality Measures

Hirota and Pedrycz [15] suggested the following five measures: validity, generality, usefulness, novelty and simplicity. These measures are adjusted in [46] for automatically mining summaries from the data in order to convert them into the fuzzy IF-THEN rules (generating IF-THEN rules from LSs with restriction part is further examined in Sect. 3.7.2).

Validity
The summary from data must be derived with high degree of truth value. The validity (v) increases, when tuples, which (partially or fully) meet the restriction also (partially or fully) meet the summarizer. The proportion of tuples, which meet S and are in R

is also called subsethood measure [30]. For the validity measure we can use either LS's truth value (validity) (3.8), or the subsethood measure.

Generality
It is essential, that reasonable amount of data supports the summary. This feature should be considered by two perspectives. First, if a data set is a small and nonrepresentative sample, then even if 99 % of data meet the summary, the result is not representative. Second, if the data set is sufficiently large (e.g. instead of a sample, the whole population of interest is analysed), then high validity could mislead in cases, when only few tuples (e.g. 2 tuples of 200 000) fully (or partially) meet the restriction part and the same tuples meet the summarizer part in (3.8)). Concerning the latter, generality is expressed as coverage [46]. For the LS with restriction, the following coverage index can be used:

$$i_c = \frac{\sum_{i=1}^{n} t(\mu_S(x_i), \mu_R(x_i))}{n} \tag{3.11}$$

where n is the number of tuples in a database. Other variables have the same meaning as in (3.8). When both $\mu_R(x_i)$ and $\mu_S(x_i)$ are included, the coverage explains how many membership degrees influence the validity of a LS. In practice, the value of coverage index is a small number, because LS with restriction usually covers a relatively small subset of the considered data. Therefore, the mapping which converts i_c into the degree of coverage suggested in, e.g. [46] can be used

$$C = f(i_c) = \begin{cases} 0, & \text{for } i_c \leq r_1 \\ 2(\frac{i_c - r_1}{r_2 - r_1})^2, & \text{for } r_1 < i_c \leq \frac{r_1 + r_2}{2} \\ 1 - 2(\frac{r_2 - i_c}{r_2 - r_1})^2, & \text{for } \frac{r_1 + r_2}{2} < i_c < r_2 \\ 1, & \text{for } i_c \geq r_2 \end{cases} \tag{3.12}$$

where $r_1 = 0.02$ and $r_2 = 0.15$. Anyway, parameters r_1 and r_2 can be set according to the user's preferences similarly as is suggested for quantifiers and predicates.

Clearly, this measure should be applied on LSs with restriction. Concerning the basic structure of LS, the validity (3.2) is affected by all tuples.

Usefulness
This feature expresses, how useful is the summary. This is also a twofold measure. From user's perspective, this measure corresponds to the goals of the task (e.g. importance of a mined summary on decision-making). Mathematically, degree of usefulness can be expressed as [46]

$$U = \min(v, C) \tag{3.13}$$

where v is proportion or validity (from the validity measure) and C is coverage of tuples (from the generality measure). Hence, usefulness can be considered as a compound measure consisting of validity and generality.

Novelty
This measure says that the unexpected summaries which meet other quality measures are very valuable for users. On the other hand, summary can be also unexpected, if it covers only outliers instead of regular behaviour in data. Hence, this summary does not representatively express relational knowledge in the data. We can say that outliers appear, if [46]

(a) the validity degree v is very small or very high
and
(b) the sufficient coverage C must be very small.
Therefore, the outliers measure can be expressed as

$$O = \begin{cases} \min(\max(v, 1-v), 1-C), & \text{for } v > 0 \\ 0, & \text{for } v = 0 \end{cases} \tag{3.14}$$

where v is the validity of a summary, C is generality (coverage) measure (3.12) and the operator *and* between (a) and (b) is expressed as the minimum t-norm (1.47). If coverage is small ($C \to 0$), then outlier measure O is near the value of 1 (if v gets value near 1 or 0). If coverage is sufficiently high (($1 - C) \to 0$), then the outlier measure is near the value of 0. The closer O is to the value of 1, the more LS is judged as sentence expressing outliers. Therefore, when $1 - O$ increases, the quality of summary increases.

Simplicity
This measure concerns the syntactic and semantic complexity of the summaries. Too complex summaries are less legible for users and require additional computational effort. The simplicity of summary expresses, how many attributes are in restriction and summarizer parts. In terms of rule bases this measure expresses, how many attributes are in antecedent and consequent parts of the rule. Therefore, simplicity can be expressed as [46]

$$S_{im} = 2^{(2-l)} \tag{3.15}$$

where l is total number of antecedents and consequences. Evidently, $S_{im} \in (0, 1]$. As the number of antecedents or consequences increases, S_{im} decreases. Obviously, this measure gets value 1 when one attribute is in R and one attribute in S parts of LS.

Generally, the LS credibly expresses the mined knowledge, when all measures are near the value of 1. But on the other hand, the low value of the simplicity measure could mean that the complex summary better explains the data. This holds for example in adding attributes into the R part of LS for focusing on a very specific part of database.

For evaluating quality for a particular summary (consisted of summarizer and restriction), measures of validity, coverage and novelty could be sufficient. This statement is explained on the interface shown in the Appendix B.

Castillo-Ortega et al. [10] examined the following four quality measures: coverage, brevity, specificity and accuracy. At first glance, first two measures are equivalent with two ones discussed in [15] and [46]. But, they are defined from different perspective of data summarization.

Coverage

The coverage C is the extent to which all tuples in a database DS are considered by $ls_i \in \mathbb{LS}$, where $\mathbb{LS}$ is the space of possible summaries. Without loss of generality, C is a normalized measure on ds_s ($ds_s = \{r_i \in DS | r_i$ is considered in $ls\}$) defined as [10]

$$\begin{cases} \text{if } ds_s = 0, & \text{then } C(ls) = 0 \\ \text{if } ds_s = DS, & \text{then } C(ls) = 1 \\ \forall ls_1, ls_2, & ds_1 \subseteq ds_2 \Rightarrow C(ls_1) \leq C(ls_2) \end{cases} \qquad (3.16)$$

Evidently, coverages (3.12) and (3.16) have different meanings and therefore value of 1 is met for different number of tuples affected in LS. This measure is suitable for summaries without quantifiers.

Example 3.5 Two illustrative summaries on a database of 2924 municipalities are
ls_1: *there are 21 municipalities with altitude over 850 m, 1446 municipalities with altitude in interval [250, 850] and 1457 below 250*
ls_2: *there are 1047 municipalities with altitude over 600 m, and 241 municipalities with altitude below 200 m.*
The coverages of these summaries are $C(ls_1) = 1$ and $C(ls_2) = \frac{1288}{2924} = 0.4405$. □

Furthermore, this approach considers only included tuples, so the crisp cardinality is used in calculation of coverage, whereas approach (3.11) considers intensities of included tuples, so membership degrees are used for computing coverage.

Brevity or shortness

This quality measure of summary ($S(ls)$) is the extent to which a summary is short, i.e. $S(ls) \in [0, 1]$ [10]. Value of 1 is reserved for the shortest possible summary. In terms of propositions included in LSs, the shortest possible summary contains only a single proposition. Regarding summaries in Example 3.5, it is clear that $S(ls_1) < S(ls_2)$.

Measure of shortness in (3.15) is influenced by number of attributes in R and S in the summary. Therefore, exponent is $2 - l$. Analogously, in order to reach value of 1 in summaries of types shown in Example 3.5 the brevity is expressed as

$$S_{im} = 2^{(1-l)} \qquad (3.17)$$

where l is total number of propositions.

Hence, the shortness measures of two summaries created in Example 3.5 are $S(ls_1) = 2^{-2} = 0.25$ and $S(ls_2) = 2^{-1} = 0.5$. An example of LS with $S = 1$ is *all municipalities have ratio of public greenery lower than 25%.*

Specificity

Measure of specificity of LS ($E_c(ls)$) is the extent to which concepts in the summary clearly identify considered data, $E_c(ls) \in [0, 1]$ [10]. $E_c(ls) = 1$ means that concerned data set DS is explained without doubt. The higher is the specificity, better is the summary, but on the other hand, length might increase.

Example 3.6 Consider two summaries from Example 3.5 and the additional one:
ls_3: *there are 1047 municipalities with altitude over 600 m, 1072 with altitude from 200 to 400 m*
Then ls_3 is more specific than ls_2 but covers less data than ls_1. □

Accuracy

This measure expresses the extent to which a summary says is true for all covered tuples in DS. $A_c(ls) = 1$ means that the summary holds for all tuples. Clearly, the higher accuracy, the better LS.

Example 3.7 Consider two summaries from 3.5, one from Example 3.6 and the additional one
ls_4: *one-third of municipalities does not exceed altitude of 192 m and two-third of municipalities do not go below this altitude*
The accuracy of ls_4 is lower than the accuracy of ls_1 but the coverage for both is equal to 1. □

3.6.2 Aggregation of Quality Measures

Naturally, the question which LS is better in terms of all measures appears. It holds for both approaches examined in Sect. 3.6.1. Although these approaches differ in definitions and types of considered LSs, all measures get values from the unit interval. Therefore, the same structure of ordering relation can be used and moreover, for all other approaches expressing quality measures in unit interval (and when is possible to convert values of measures to this interval).

The basic quality ordering relation $\leq_Q$ (*having less quality than*) is defined as a binary relation on $\mathbb{LS}$ [10]

$$\forall ls_1, ls_2 \in \mathbb{LS} \; ls_1 \leq_Q ls_2 \Leftrightarrow C(ls_1) \leq C(ls_2) \land S_{im}(ls_1) \leq S_{im}(ls_2) \land \\ E_c(ls_1) \leq E_c(ls_2) \land A_c(ls_1) \leq A_c(ls_2) \tag{3.18}$$

This measure generally works when ls_i is better than ls_j by all measures. Otherwise, it is not an easy task, because measures might be conflicting or partially redundant. Apparently, creating generalized aggregation function covering quality measures is a challenge.

Quality measures discussed above can be calculated automatically from the database and obtained summaries. On the other side, quality measures of approach created in [38] are based on experts that analyse obtained summaries. This approach is suitable for summarizing data expressed as numbers and pictures. The quality measures are evaluated as values in the [0, 5] or [0, 10] interval. Anyway, these degrees can be converted to the unit interval, if quality ordering relation such as (3.18) need to be applied.

Summing up, a quality criterion can contain measures of both types: objective (automatically calculated) and subjective (evaluated by expert(s)). Issues like conflicting measures, vague measures and partially redundant ones should not be neglected.

3.6.3   Influence of Constructed Fuzzy Sets and T-Norms on Quality

In this section, impact of chosen t-norm on the validity is examined, together with influence of constructed fuzzy sets on the coverage and validity.

For the LSs of structure Q R are S the quality is measured for each data point x_i ($i = 1, ..., n$) by t-norm in the numerator of (3.8) [31]. This is one aspect of the complex problem of quality. Chosen t-norm influences not only aggregating restriction and summarizer, but also the truth value of conjunction between the atomic predicates inside restriction and summarizer. T-norms meet all axiomatic properties, but differ in satisfying algebraic properties (Sect. 1.3.1).

For instance, when each atomic predicate P_j ($j = 1, ..., m$) in summarizer is satisfied with degree of 0.47, then the tuple should participate in S with the degree of 0.47. Only minimum t-norm (1.47) meets this requirement. Furthermore, this t-norm is not nilpotent and does not have limit property. Łukasiewicz t-norm (1.49) meets the nilpotency property, causing that tuple participates in proportion with value of 0. Product t-norm (1.48) meets the limit property, causes that participation of tuples is decreasing in the proportion, in which the number of atomic predicates increases. When $j = 1$ tuple participates with degree of 0.47, but when $j = 5$, the same tuple participates in summary with 0.02293.

The only suitable t-norm for merging conjunction of atomic conditions in R and S parts of LSs is the minimum one, because it does not unnaturally reduce the proportion of tuples in a data set that satisfy LS. An appropriate t-norm influences the quality of mined LSs, but further quality aspects should be included.

Fuzzy sets allow users to express uncertainty related to linguistic summaries. But, the subjectivity in constructing fuzzy sets may influence quality of summarized information. It especially holds for the sufficient data coverage and outliers.

As was already mentioned in queries, the domains of attributes are, during the database design phase, defined in a way that all theoretically possible values can be stored. However, in practice, collected data fill only parts of respective domains. Hence, the situation plotted in Fig. 3.5, where L and H are the lowest and the highest values in the current content of attributes respectively, and D_{min} and D_{max} are the

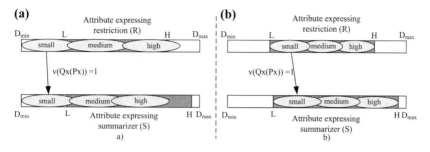

Fig. 3.5 Constructed fuzzy sets for restriction and summarizer. **a** fuzzy sets do not reflect reality (stored data); **b** fuzzy sets reflect stored data

lower and upper limit of domains respectively, might appear. The truth value equal to 1 in Fig. 3.5a expresses summary from the outliers and therefore is of low quality.

In order to mitigate this problem, membership functions should consider only parts of domains that contain data. The validity equal to 1 in Fig. 3.5b can be the summary of good quality. But the data distribution far from the uniform one might cause that mined LS expresses relations on insufficient data coverage. Let us imagine that only 30 of $5 \cdot 10^7$ tuples fully meet the R and the same tuples more or less fully meet the S, then the validity equal to 1 leads to the irrelevant linguistic interpretation of data.

This problem can be solved by aggregating validity and coverage. In general, the outlier measure (3.14) for $v > 0$ can be expressed as

$$O = t(s(v, 1 - v), (1 - C)) \tag{3.19}$$

where t is t-norm, s is t-conorm and C is coverage (3.12).

The non-outlier measure, i.e. the negation of (3.19), is obtained by De Morgan's law as

$$1 - O = s(t(v, 1 - v), C) \tag{3.20}$$

We can say that LS is of high quality Q_c, if both validity and non-outliers measure are high. This observation can be expressed as

$$Q_c = t(v, 1 - O) = t(v, s(t(v, 1 - v), C)) \tag{3.21}$$

If we define coverage as significant, when its value is higher or equal 0.5 and use the property of t-norms $t(1 - v, v) \le 0.5$, then from (3.21) yields

$$Q_c = \begin{cases} t(v, C), & \text{for } C \ge 0.5 \\ 0, & \text{otherwise} \end{cases} \tag{3.22}$$

Obviously, the question, which t-norm to use, appears again. Let us have for illustrative purpose, values of validity and coverage for two LSs shown in Table 3.5.

Table 3.5 Aggregating validity and coverage by t-norms

Summary	Validity	Coverage	Q_c by min t-norm	Q_c by product t-norm
ls 1	0.75	0.75	0.75	0.5625
ls 2	0.75	0.97	0.75	0.7275

Table 3.6 Quality of mined LSs

Summary	Validity	Coverage	Q_c
ls 1	0.93	1	0.93
ls 2	0.34	0.78	0.2652
ls 3	0.75	0.75	0.5625
ls 4	1	0.67	0.67
ls 5	0.91	0.71	0.6461
ls 6	0.83	0.34	0

The minimum t-norm says that *ls 1* and *ls 2* are undistinguishable, because it does not consider degrees greater than the minimal one. The same holds for the nilpotent minimum t-norm. The suitable t-norms of the basic ones are product and Łukasiewicz t-norms. It is an observation opposite to that of aggregating atomic predicates inside S and R.

Example 3.8 In order to mine all relevant summaries, the user has defined fuzzy sets for attributes appearing in restriction and summarizer and assigned parameters for quantifiers. For illustrative purpose, mined LSs are written as *ls i* ($i = 1, \ldots, 6$) and shown in Table 3.6.

The aggregation of validity and coverage by (3.22) says that the summary of the highest quality is *ls 1*. Although *ls 4* has the maximal value of validity, its coverage is significantly lower than data coverage in summary *ls 1*. Summary *ls 6* is excluded from the set of relevant summaries due to low coverage, although its validity is significant.   □

3.7 Some Applicability of LS

Linguistic summaries could be applied in a variety of tasks. Due to limited scope of this book, only several possible applications are discussed. Other applications dealt with expressing trends in time series [27, 28]; bipolar linguistic summaries [12]; interval-valued linguistic summaries [37] and the like. LSs on fuzzy data are examined in Sect. 5.7.

3.7.1 Quantified Queries (LS as a Nested Condition)

This is a class of database queries which use linguistic quantifiers as query conditions. This class is especially suitable for the 1:N relationships in a relational database such as: REGION-MUNICIPALITY and CUSTOMER-INVOICE. This relationship means that one region contains more municipalities and one municipality belongs to the exactly one region. An example of a quantified query might be as follows: *select regions where most of municipalities have small water consumption per inhabitant.* In the first step, the validity of summaries is calculated for each region based on the municipalities data. In the second step, regions are ranked downwards starting with region having the highest value of validity. The procedure for calculating validities is straightforwardly created as the extension of (3.2) in the following way [17]:

$$v_j(LS) = \mu_Q(\frac{1}{N_j}\Sigma_{i=1}^{N_j}\mu_S(x_{ij})) \quad j = 1...C \quad \Sigma_{j=1}^{C}N_j = n \qquad (3.23)$$

where n is the number of entities in the whole database, N_j is the number of entities in group j (e.g. municipalities belonging to the region j), C is the number of groups in database (e.g. regions), v_j is validity of LS for j-th group and $\frac{1}{N_j}\Sigma_{i=1}^{N_j}\mu_S(x_{ij})$ is the proportion of tuples in j-th group that satisfies summarizer S.

Although the summarization is focused on the part of database, we are working with the basic structure of LS, because database is divided into logically and sharply separated parts (in our case regions).

When the quantified nested query condition contains LSs with restriction (3.8) such as *select customers where most of small volume of goods in orders has high price*, the query procedure for calculating validities can be straightforwardly extended to

$$v_j(LS) = \mu_Q(\frac{\Sigma_{i=1}^{N_j}f(\mu_S(x_{ij}), \mu_R(x_{ij}))}{\Sigma_{i=1}^{N_j}\mu_R(x_{ij})}) \quad j = 1...C \quad \Sigma_{j=1}^{C}N_j = n \qquad (3.24)$$

where variables have the same meaning as in (3.8) and (3.23).

Possibilities of LSs in quantified queries are demonstrated in the next three examples to illustrate variety of possible queries and applications.

Example 3.9 An agency realizing surveys by questionnaires would like to know, which questionnaires are most demanding for respondents. The query is *select questionnaires where most of respondents have high response time.* The response times for questionnaire Qu_1 are in Table 3.7. In the same way, response times for other questionnaires are recorded.

In the first step, fuzzy sets *small*, *medium* and *high* are constructed on the domain [10, 50] by the uniform domain covering method keeping the relation $\varepsilon = 2\theta$ (Fig. 2.2), because the fastest answer is reached in 10 min and the slowest in 50 min. The quantifier *most of* has parameters $m = 0.5$ and $n = 0.85$ (Fig. 3.1).

Table 3.7 Time required for filling questionnaire Qu_1

Respondent	Time (min)
Resp 1	36
Resp 2	21
Resp 3	40
Resp 4	39
Resp 5	50
Resp 6	38
Resp 7	44
Resp 8	10
Resp 9	37
Resp 10	46

Table 3.8 Selected questionnaires by quantified query

Questionnaire	Matching degree to query
Qu_3	0.8925
Qu_1	0.2857

Questionnaires may be of different complexity. Thus, these fuzzy sets should be constructed for each questionnaire according to the response times.

Applying the basic structure of LSs, the validity 0.2857 is reached. Let us think that in the same way validities of Qu_2 and Qu_3 are 0 and 0.8925, respectively. The result of query is shown in Table 3.8, where matching degree of the questionnaire to query is validity of quantified question.

The solution says that Qu_3 should be re-designed in order to decrease response burden. Otherwise, respondents may not cooperate in the future surveys or put less attention to question, which might result in lower quality of surveyed data.

This approach provides further benefit. In Table 3.7 there may be many other attributes, such as respondents' answers to each question and their personal data. This query keeps data that are sensitive or are out of interest undisclosed. Agency knows that the most problematic questionnaire is Qu_3 and therefore only this questionnaire should be considered.   □

The next example illustrates deeper dive into the hierarchical structure of nodes, in our case territorial units.

Example 3.10 An environmental agency is interested in revealing areas of small waste production. The hierarchical structure of a fictive country is shown in Fig. 3.6. The LS is *most of municipalities have small waste production per inhabitant*. Small amount of produced waste could be caused by several reasons: low income (low spending power implies low amount of goods bought and therefore smaller volume of produced waste), well-developed recycling system and education, avoiding the waste fee by burning it out in furnaces or in gardens. In order to find the right answer, further attributes should be analysed: income, amount of collected waste for recycling, pollution, number of respiratory diseases and the like.

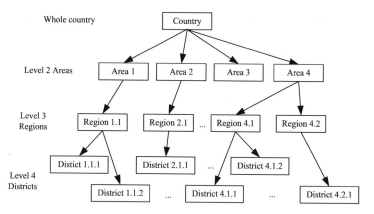

Fig. 3.6 Hierarchy of territorial units

The summarization is evaluated on the second level (Areas 1–4). In order to obtain parameters for the fuzzy set *small waste production per inhabitant* the attribute is fuzzified into three fuzzy granules (small, medium and high) uniformly distributed on the domain of recorded waste production. Let us consider that validities for each territorial unit are

- Area 1: $v(Qx(Px)) = 0.13$
- Area 2: $v(Qx(Px)) = 0$
- Area 3: $v(Qx(Px)) = 0.34$
- Area 4: $v(Qx(Px)) = 0.98$

According to these validities, the attention should be focused on the area number 4. This level contains two regions: Region 4.1 and Region 4.2. The same LS (with adjusted parameters to respective domains) on this level computed the following:

- Region 4.1: $v(Qx(Px)) = 1$
- Region 4.2: $v(Qx(Px)) = 0.827$

If we want to go deeper, the same LS can be employed on all districts in Region 4.1 and all districts in Region 4.2 (because both regions bear significant validities of the LS). For each district with high validity of the LS, further analysis of indicators such as income, developed recycling systems and education, health conditions could reveal the main reasons for the low waste production. Reasons for small waste production may be different in particular districts. □

The final example in this section is focused on quantified queries merging constraints and wishes. This approach can be solved by bipolar approaches [22] and asymmetric conjunction. The latter is shown in the following example:

Table 3.9 Matching degrees to atomic predicates, constraints, wishes and to overall query condition

Village	P_1	P_2	P_3	P_4	P_5	P_6	P_7	μ_{Q_C}	μ_{Q_W}	α
Vil 1	0.8	0.9	0.6	0	1	0.7	0.75	0.21	0.89	0.21
Vil 2	0	0	0	0.3	0.2	0	0	0	0	0
Vil 3	1	0	1	1	1	1	1	0.71	1	0.855
Vil 4	0.2	0	0.4	0	1	1	0.75	0	1	0
Vil 5	0.9	0.9	0.8	1	0	0	0.25	1	0	0.5
Vil 6	0.9	0.9	0	1	0.6	0.8	0.75	0.57	0.61	0.59

Example 3.11 This example is continuation of Example 2.10. The task is to find suitable village for building house. Let the following be the relevant predicates for evaluating villages: altitude above sea level around 1500 m (P_1), small population density (P_2), medium area of village size (P_3), low pollution (P_4), high number of sunny days (P_5), short distance to the region capital (P_6) and positive reviews about public transport (P_7). It is highly presumable that none of villages meets all predicates in a query of the structure $\bigwedge_{i=1}^{7} P_i$, even though predicates have flexible boundaries.

In order to solve this problem, user may say that village should be considered, if it meets most of predicates. Hence, query matching degree for tuple r is calculated by $\mu_Q(r) = \mu_Q(\frac{\sum_{i=1}^{7} \mu_{P_i}(r)}{7})$.

Furthermore, not all predicates may be equally important. Let us say that P_1, P_2, P_3 and P_4 are constraints and P_5, P_6 and P_7 are wishes.

The matching degrees of villages to respective predicates are shown in Table 3.9.

Results are obtained in the following way (using quantifier *most of* expressed by parameters $m = 0.5$ and $n = 0.85$):

- tuple's matching degree to constraints: $\mu_C(r) = \mu_{Q_C}(\frac{\sum_{i=1}^{4} \mu_{C_i}(r)}{4})$
- tuple's matching degree to wishes: $\mu_W(r) = \mu_{Q_W}(\frac{\sum_{j=1}^{3} \mu_{C_j}(r)}{3})$
- tuple's overall matching degree applying (2.25): $\alpha = \min(\mu_C, \frac{\mu_C + \mu_W}{2})$

It is obvious from Table 3.9 that the *and* connective results in an empty answer, because the value of 0 is the annihilator for conjunction calculated by any t-norm.

When quantified constraint is fully rejected, then the solution is 0, because the influence of the wish part in this case is irrelevant (the case of (*Vil 4*)). But, when quantified wish is not met, then the solution is lower than in case, when only constraint is considered and higher when *and* connective is applied (*Vil 5*). High values of constraints allow wishes to influence solution. Thus, the best option is *Vil 3* followed by *Vil 6* and *Vil 5*.   □

3.7.2 Generating IF-THEN Rules

Sophisticated and powerful approaches, such as neural networks and genetic algorithms, are widely used in learning rules from the data. If a small business or agency
wants to apply flexible rules to managerial and other decisions, then complex tools
are usually beyond their resources. Contrary, mining LSs from the data does not
require complex tools.

The IF-THEN rules can be generated from the LSs with the restriction part. LSs,
which meet quality requirements (Sect. 3.6) can be transferred into the (weighted)
fuzzy IF-THEN rules [46]. Only single-antecedent (one attribute in the restriction part
of LS) and single-consequent (one attribute in the summarizer part of LS) rules are
examined in this section. Multi-antecedent and multi-consequent rules are discussed
in [46].

Generally, a weighted fuzzy rule merges input linguistic terms with the output
terms (classes) in the following way [8]:

$$\text{if } x_i \text{ is } A_i^r \text{ then class is } C^r \text{ with } cf = \varphi^r \qquad (3.25)$$

where x_i is the i-th input, A_i^r is the membership function for i-th input in r-th rule,
C^r denotes the output class, cf is a confidence factor representing the rule certainty
(or validity) and $\varphi^r \in [0, 1]$ is value of cf for a rule r.

The canonical form of LS can be expressed as IF-THEN rule [46]

$$\text{if } A_1 \text{ has } R \text{ then } A_2 \text{ is/has } S \, [Q] \qquad (3.26)$$

where R is restriction, S is summarizer of LS (3.8) and $[Q] \in [0, 1]$ is quality
measure indicating, how good is the rule. The measure $[Q]$ is equivalent with the
weight in (3.25), when weight is based on the quality measures. Generally speaking,
both measures express doubt about a rule which is further used in inference systems.

For a given data set user should specify the terms used for each antecedent (R)
and consequent (S) from the term sets $\overline{R}$ and $\overline{S}$, respectively (3.10). Concerning
the term set for quantifiers $\overline{Q}$, only terms such as *most of* or *almost all* are suitable.
Quantifiers such as *few* are not suitable in tasks of generating rules. The next step (the
main challenge) is to compute $[Q]$, which should merge measures listed in Sect. 3.6.

Example 3.12 The aim of this example is to illustrate generating rules from the LSs.
In Example B.3, we were interested in validity of summary *most of municipalities
with high ratio of arable land have small altitude above sea level.*

The validity v is high (0.814). Concerning generality or coverage C (3.12), value
is 0.9927. Regarding usefulness, user can say that this rule is useful for the purpose of constructing rule base. The outlier measure (3.14) is 0.0047, i.e. this quality
measure gets value of 0.9927. Finally, simplicity of rule is equal to 1. This LS is of
high quality and therefore it is converted to rule. For aggregating atomic measures,
product t-norm is used. Hence, the quality measure is $[Q] = 0.8081$. Rule is of the
structure (3.26)

if ratio of arable land is high, then altitude above sea level is small [0.8081]

The same rule can be expressed as weighted one of the structure (3.25)

if ratio of arable land is high, then altitude above sea level is small with
cf = 0.8081   □

For the sake of simplicity, if all other quality measures are met, validity can be used as weight of the rule.

In Example 3.2 we were interested in validity of summary *most of expensive books have small number of pages*. The validity of LS was high and therefore may be suitable for converting into the rule. Let us look at other quality measures. The coverage index i_c is 0.27, which means that coverage C gets value of 1. The novelty measure is high: mathematically, it is not based on outliers, because O is equal to 0 and we have learned something, which was not obvious. Finally, simplicity measure S is met with maximal value. Hence, $cf = 1$.

In Example 3.4 validities of relevant LSs have been calculated (Table 3.4). For the simplicity, let be all quality measures fully met, then from the Table 3.4 and (3.25) or (3.26), the following two rules are created:

• *if Population density is small, then Production of waste is small [1]*
• *if Population density is high then Production of waste is high [0.662]*

The first LS is undoubtedly true and therefore $Q = 1$ or $cf = 1$. The second LS is more or less significantly satisfied and converted into rule with lower value of quality measure (weight). The third row in the Table 3.4 cannot be converted to the IF-THEN rule, because the validity is insignificant. The weight may be further considered in intensity of rule firing.

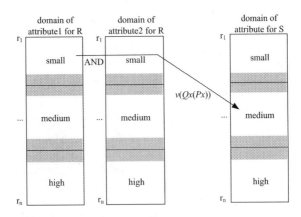

Fig. 3.7 LS with a stronger restriction part consisted of two attributes merged with the *and* operator

3.7.3 Estimation of Missing Values

This problem appears, when values of attributes are not known for some tuples. Missing values are due to the fact that data is not available because of several reasons (e.g. nonavailability of instruments to measure phenomena in all units and reluctance of respondents to cooperate in surveys) [16]. In case of database queries, we are not sure, whether a nonselected tuple is far to meet the query condition, or because the value of the attribute is missing. In classification, tuples with missing values cannot be properly classified. This issue also influences construction aggregates consisted of several attributes. For example, nonresponded data and errors in case of statistical data collection are far to be negligible [4].

For instance, in the databases of territorial units' statistics we could recognize some similarities between units (e.g. climatic conditions and water consumption or population density and waste production). We could presume, that for example territorial units, which have similar values of population density, income and ratio of the built-up area, have also similar waste production. This hypothesis could be validated by the LSs. If validity is high then, the waste production for units, where this indicator is missing, can be roughly estimated. For this purpose we need to calculate validity (3.8) by restrictive quantifier. The solution is the quantifier *almost all*.

Further, when validity is significant, but not sufficiently strong or quality measures are not sufficiently high, presumably in a more restricted part of database the dependency is stronger. In order to reveal, whether this assumption is true, two options can be applied. The first option is the conjunction of initial and adjacent atomic predicates in the R part, in order to focus on more restricted part, where we expect stronger relation. This case is illustrated in Fig. 3.7. The second option is finer granulation of attributes' domains depicted in Fig. 3.8.

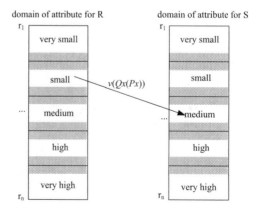

Fig. 3.8 LS with a finer granulation of attributes domains

Regarding quality measures, validity and coverage are more important than simplicity. More restrictive part of database may have strong relation between attributes (high values of validity and coverage), but the simplicity measure (3.15) is not high. From the bipolar perspective, we can say that validity and coverage are restrictions and simplicity is wish.

Example 3.13 Let us assume that values of attribute A_p are missing for some municipalities. These territorial units cannot be properly classified for business or policy decision-making. Hence, these values should be at least roughly estimated. We presume that territorial units, which have similar values of attribute A_1 have also similar values of A_p. Let us estimate only LS of structure *almost all municipalities having small value of A_1 have small value of A_p* has high validity and high values of quality measures. Let us state the value for territorial unit M is missing.

In the second step, value of A_1 for M ($val_{A_1}(r_m)$) is retrieved from database. If this value belongs to the *small value of A_1*, the procedure may proceed. Fuzzy query for retrieving similar tuples (Example A.2) to tuple r_m has the following structure:

SELECT Ap
FROM municipalities
WHERE A1 is about $val_{A_1}(r_m)$

The answer is vector of values of attribute A_p. The estimated value is computed from this vector. It can be calculated as usual average of selected tuples from database or as average influenced by membership degrees.

If LSs focused on some parts of A_1 and Ap domains are of insufficient quality, additional attribute(s), which is (are) assumed to be related to A_1 and Ap is (are) included. The stronger R (atomic predicates merged by the *and* connective) may reveal that a very significant relation occurs in one part of the database. But the number of combinations increases. Techniques for optimization and excluding LSs of possible low validity should be used.   □

For instance, in the official statistics the data imputation (estimation of missing values) is a topic of significant interest. One of used methods is the Hot Deck imputation, due to its simplicity and good results. Each missing value is replaced with data from a more or less similar tuple using the linear restriction rules [11]. Hot Deck is efficiently used in practice, even though its theory is not as well developed as in other methods [2]. Neural networks [20] and genetic algorithms [29] (soft computing approaches) are also very valuable and powerful tools in this direction, though somewhat complex. Their advantage is in estimating values of datasets containing indicators which are complexly influenced, e.g. export of similar goods might significantly vary in some periods, due to variety of internal and external influences. On the other hand, LSs are not so complex and able to process nonlinear and flexible, but relatively stable relations. For example, if similar territorial units have almost the same climatic conditions, then it is expectable that in other periods (for which we do not have data for some of the units) they will have also similar values for attributes under interest.

3.8 Building Summaries

When developing software solutions, we should consider simplicity for users and modular extensions of future needs in the similar way, as was discussed in Sect. 2.6. For instance, the interface of LSs with restriction (Fig. B.2) is straightforwardly extended with quality measures (Fig. B.3), when required. It may be useful to store LSs, which are frequently used and then slightly adjust them, if needed. In order to store fuzzy sets appearing in summaries and IF-THEN rules, relational databases are an option, because required data are often stored in relational databases. A possible option is the fuzzy meta model elaborated in Sect. 5.5.

It is not necessary to build additional functionalities in the software tool from the beginning.

> Fuzzy logic is flexible, that is; with any given system, it is easy to layer on more functionality without starting again from scratch [35].

If company already has tool for fuzzy queries, it can be extended with LSs capabilities. Interfaces and application related to summarizing territorial units are adjusted from fuzzy querying interfaces shown in Appendix A and illustrated in Appendix B.

The architecture plotted in the Fig. 2.8 may be the basis for tool building for LSs. The parts related to the LSs (new interface and adjusted procedures) can be added. The architecture enveloping fuzzy queries and LSs is shown in Fig. 3.9. In case of basic LS, user is able to define parameters or ask for suggestion by the same procedure as in Example A.5. In case of LS with restriction, parameters are mined from the current content of database and suggested in the same way as in Example B.2.

Roughly speaking, two general types of applications and interfaces can be applied. The first type is related to mining validity of a specific LS of interest or for all relevant ones. The second type is related to quantified query conditions for data mainly stored in relations merged by the 1:N relationships. In the latter, an interface managing such LSs should contain module for presenting selected tuples in useful and understandable way, as was discussed in Appendix A.

Procedures for LSs

In this section, the algorithms for three cases: evaluation of single LS, automatic generation of linguistic summaries from predefined term sets and for rule generation are outlined.

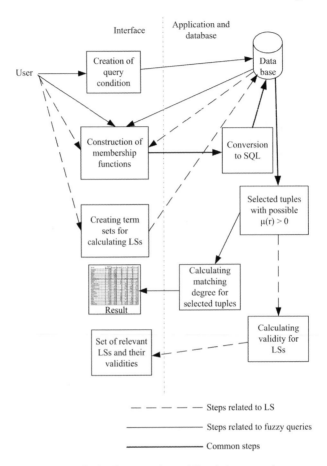

Fig. 3.9 An architecture enveloping fuzzy queries and linguistic summaries

Single LS

1. Check which attributes and fuzzy sets have been selected by the user for R and S and which quantifier is chosen
2. Open database connection
3. Select the smallest and the highest values of attributes appearing in summary by SQL query
4. Calculate parameters for chosen fuzzy sets for all attributes
5. Offer these parameters to user for acceptance or modification
6. Select all tuples from database, which have $\mu_S(x) > 0$ and $\mu_R(x) > 0$
7. Close database connection
8. Calculate proportion of entities, validity v and quality measures of LS
9. Present solution to user

Automatic generation of linguistic summaries

1. Check which attributes have been chosen by the user
2. Create all possible combinations (n) of LSs
3. Open database connection
4. Select the smallest and the highest values by SQL query for all attributes
5. Calculate parameters for all fuzzy sets of R and S
6. Apply all required quantifiers
7. For $i = 1$ to n

   7.1 Select all tuples from database, which have $\mu_{S_i}(x) > 0$ and $\mu_{R_i}(x) > 0$
   7.2 Calculate proportion of entities, validity and quality measures of LS

8. Close database connection
9. Present solution to user in a tabular form

From linguistic summaries to rules

This procedure is the same as the previous except slight modification of steps 1 and 7.2.

1. Check which attributes, quantifiers, linguistic terms and threshold value β have been chosen by the user
7. 2. Calculate proportion of entities and validity of LS

- If validity is greater than β, then calculate other quality measures
  If quality measures are sufficiently high, then LS is converted into rule
- Else LS is not accepted as rule

The advantage of LSs is in low requirements for the software tools. Nowadays, majority of companies use relational databases. By the SQL-like query we can retrieve all relevant information. In this chapter, tuples involved in summaries are selected by the query approach based on the GLC (2.9). Other fuzzy query approaches provide space for working with summaries like FQUERY [25], SAINTETIQ [39, 45], SQLf [5] and SummarySQL [41].

References

1. Almeida, R.J., Lesot, M-J., Bouchon-Meunier, B., Kaymak, U., Moyse, G.: Linguistic summaries of categorical time series patient data. In: 2013 IEEE International Conference on Fuzzy Systems (FUZZ-IEEE 2013), pp. 1–8, Hyderabad (2013)
2. Andridge, R., Little, R.: A review of Hot Deck imputation for survey non-response. Int. Stat. Rev. **78**, 40–64 (2010)
3. Arguelles, L., Triviño, G.: I-struve: automatic linguistic descriptions of visual double stars. Eng. Appl. Artif. Intell. **26**, 2083–2092 (2013)
4. Bavdaž, M.: Sources of measurement errors in business surveys. J. Official Stat. **26**, 25–42 (2010)
5. Benali-Sougui, I., Sassi-Hidri, M., Grissa-Touzi, A.: Flexible SQLf query based on fuzzy linguistic summaries. In: International Conference on Control, Engineering and Information Technology (CEIT 2013), vol. 1, pp. 175–180, Sousse (2013)

6. Bosc, P., Dubois, D., Pivert, O., Prade, H., De Calmes, M.: Fuzzy summarization of data using fuzzy cardinalities. In: XXth International Conference on Information Processing and Management of Uncertainty in Knowledge-Based Systems (IPMU 2002), pp. 1553–1559, Annecy (2002)
7. Bouchon-Meunier, B., Moyse, G.: Fuzzy linguistic summaries: where are we, where can we go? In: IEEE Conference on Computational Intelligence for Financial Engineering & Economics, pp. 1–8, New York (2012)
8. Branco, A., Evsukoff, A., Ebecken, N.: Generating fuzzy queries from weighted fuzzy classifier rules. In: ICDM workshop on Computational Intelligence in Data Mining, pp. 21–28, Houston (2005)
9. Campbell, M.J., Swinscow, T.D.V.: Statistics at Square One, 11th edn. Wiley, West Sussex (2009)
10. Castillo-Ortega, R., Marín, N., Sánchez, D., Tettamanzi, A.: Quality assessment in linguistic summaries of data. In: 14th International Conference on Information Processing and Management of Uncertainty in Knowledge-Based Systems (IPMU 2012), pp. 285–294, Catania (2012)
11. Coutinho, W., de Waal, T.: Hot Deck imputation of numerical data under edit restrictions. Discussion paper 2012/23, Statistics Netherlands, The Hague/Heerlen (2012)
12. Dziedzic, M., Kacprzyk, J., Zadrożny, S.: On some quality criteria of bipolar linguistic summaries. In: Federated Conference on Computer Science and Information Systems (FedCSIS 2013), pp. 643–646, Cracow (2013)
13. George, R., Srikanth, R.: Data summarization using genetic algorithms and fuzzy logic. In: Herrera, F., Verdegay, J.L. (eds.) Genetic Algorithms and Soft Computing, pp. 599–611. PhysicaVerlag, Heidelberg (1996)
14. Glöckner, I.: Quantifier selection for linguistic data summarization. In: 2006 IEEE International Conference on Fuzzy Systems, pp. 720–727, Vancouver (2006)
15. Hirota, K., Pedrycz, W.: Fuzzy computing for data mining. Proc. IEEE **87**, 1575–1600 (1999)
16. Hudec, M.: Linguistic summaries applied on statistics—case of municipal statistics. Austrian J. Stat. **43**, 63–75 (2014)
17. Hudec, M.: Issues in construction of linguistic summaries. In: Mesiar, R., Bacigál, T. (eds.) Proceedings of Uncertainty Modelling 2013, pp. 35–44. STU, Bratislava (2013)
18. Hudec, M., Vučetić, M., Vujošević, M.: Synergy of linguistic summaries and fuzzy functional dependencies for mining knowledge in the data. In: 18th IEEE International Conference on System Theory, Control and Computing (ICSTCC 2014), pp. 335–340, Sinaia (2014)
19. Hudec, M., Vučetić, M., Vujošević, M.: Comparison of linguistic summaries and fuzzy functional dependencies related to data mining. In: Alam, S., Dobbie, G., Sing Koh, Y., ur Rehman, S. (eds.) Biologically-Inspired Techniques for Knowledge Discovery and Data Mining, pp. 174–203. Information Science Reference, Hershey (2014)
20. Juriová, J.: Use of neural networks for data mining in official statistics. In: New Techniques and Technologies for Statistics (NTTS 2011), Brussels (2011)
21. Kacprzyk, J., Yager, R.R.: Linguistic summaries of data using fuzzy logic. Int. J. Gen. Syst. **30**, 33–154 (2001)
22. Kacprzyk, J., Zadrożny, S.: Compound bipolar queries: combining bipolar queries and queries with fuzzy linguistic quantifiers. In: 8th Conference of the European Society for Fuzzy Logic and Technology (EUSFLAT 2013), pp. 848–855, Milano (2013)
23. Kacprzyk, J., Zadrożny, S.: Protoforms of linguistic database summaries as a human consistent tool for using natural language in data mining. Int. J. Softw. Sci. Comput. Intell. **1**, 100–111 (2009)
24. Kacprzyk, J., Zadrożny, S.: Linguistic database summaries and their protoforms: towards natural language based knowledge discovery tools. Inf. Sci. **173**, 281–304 (2005)
25. Kacprzyk, J., Zadrożny, S.: Computing with words in intelligent database querying: standalone and internet-based applications. Inf. Sci. **134**, 71–109 (2001)
26. Kacprzyk, J., Wilbik, A., Zadrożny, S.: Linguistic summarization of time series using a fuzzy quantifier driven aggregation. Fuzzy Sets Syst. **159**, 1485–1499 (2008)

27. Kacprzyk, J., Wilbik, A., Zadrożny, S.: Linguistic summarization of trends: a fuzzy logic based approach. In: 11th Information Processing and Management of Uncertainty in Knowledge-based Systems (IPMU 2006), pp. 2166–2172, Paris (2006)
28. Kacprzyk, J., Wilbik, A., Zadrożny, S.: On some types of linguistic summaries of time series. In: 3rd International IEEE Conference on Intelligent Systems, pp. 1–8, New York (2012)
29. Kľučik, M.: Estimates of foreign trade using genetic programming. In: 46th Scientific Meeting of the Italian Statistical Society (SIS 2012), Rome (2012)
30. Kosko, B.: Fuzziness vs. probability. Int. J. Gen. Syst. **17**, 211–240 (1990)
31. Lesot, M-J., Moyse, G, Bouchon-Meunier, B.: Interpretability of fuzzy linguistic summaries. Fuzzy Sets Syst. **292**, 307–317 (2016)
32. Lieétard, L.: A functional interpretation of linguistic summaries of data. Inf. Sci. **188**, 1–16 (2012)
33. Liu, B.: Uncertain logic for modeling human language. J. Uncertain Syst. **5**, 3–20 (2011)
34. Mani, I., Maybury, M.T.: Advances in automatic text summarization. The MIT Press, Massachusetts (1999)
35. Meyer, A., Zimmermann, H.J.: Applications of fuzzy technology in business intelligence. Int. J. Comput. Commun. Control. V **I**(3), 428–441 (2011)
36. Moyse, G., Lesot, M-J., Bouchon-Meunier, B.: Linguistic summaries for periodicity detection based on mathematical morphology. In: 2013 IEEE Symposium on Foundations of Computational Intelligence (FOCI), pp. 106–113, Singapore (2013)
37. Niewiadomski, A., Ochelska, J., Szczepaniak, P.S.: Interval-valued linguistic summaries of databases. Control Cybern. **35**, 415–443 (2006)
38. Pereira-Fariña, M., Eciolaza, L., Triviño, G.: Quality assessment of linguistic description of data. In: ESTYLF, pp. 608–612, Valladolid (2012)
39. Raschia, G., Mouaddib, N.: SAINTETIQ: a fuzzy set-based approach to database summarization. Fuzzy Sets Syst. **129**, 137–162 (2002)
40. Rasmussen, D., Yager, R.R.: Finding fuzzy gradual and functional dependencies with Summary SQL. Fuzzy Sets Syst. **106**, 131–142 (1999)
41. Rasmussen, D., Yager, R.R.: Summary SQL—a fuzzy tool for data mining. Intell. Data. Anal. **1**, 49–58 (1997)
42. Smits, G., Pivert, O., Hadjali, A.: Fuzzy cardinalities as a basis to cooperative answering. In: Pivert, O., Zadrożny, S. (eds.) Flexible Approaches in Data, Information and Knowledge Management. Studies in Computational Intelligence, vol. 497, pp. 261–289. Springer, Berlin, Heidelberg (2013)
43. Srikanth, R., Agrawal, R.: Mining quantitative association rules in large relational databases. In: ACM SIGMOD International Conference on Management of Data, pp. 1–12, Montreal (1996)
44. van der Heide, A., Trivino, G.: Automatically generated linguistic summaries of energy consumption data. In: Intelligent System Design And Applications (ISDA 2009), pp. 553–559, Pisa (2009)
45. Voglozin, W.A., Raschia, G., Ughetto, L., Mouaddib, N.: Querying a summary of database. J. Intell. Inf. Syst. **26**, 59–73 (2006)
46. Wu, D., Mendel, J.M., Joo, J.: Linguistic summarization using if-then rules. In: 2010 IEEE International Conference on Fuzzy Systems, pp. 1–8, Barcelona (2010)
47. Yager, R.R.: A new approach to the summarization of data. Inf. Sci. **28**, 69–86 (1982)
48. Yager, R.R., Ford, M., Canas, A.J.: An approach to the linguistic summarization of data. In: 3rd International Conference of Information Processing and Management of Uncertainty in Knowledge-based Systems (IPMU 1990), pp. 456–468, Paris (1990)
49. Yu, J., Reiter, E., Hunter, J., Sripada, S.: SumTime-turbine: a knowledge-based system to communicate gas turbine time-series data. In: Chung, P.W.H., Hinde, C.J., Ali, M. (eds.) Lecture Notes in Computer Science (LNAI), vol. 2718, pp. 379–384. Springer, Berlin, Heidelberg (2003)
50. Zadrożny, S., Kacprzyk, J.: Issues in the practical use of the OWA operators in fuzzy querying. J. Intell. Inf. Syst. **33**, 307–325 (2009)

Chapter 4
Fuzzy Inference

Abstract In practice we can find many examples of inference rules, where relation between antecedent and consequence is expressed by linguistic terms, e.g. *if turnover is high, then provide high discount* in business or *if temperature is low and humidity is medium then turn valve slightly up* in controlling technical systems. Furthermore, attributes' values either measured or estimated are of both kinds: crisp and vague or fuzzy. We start by formalizing single fuzzy rule with one antecedent and finish with formalizing multiple fuzzy rules containing several antecedents. Both models of fuzzy inference (Mamdani and Sugeno) are examined. The solution depends on chosen fuzzy sets, logical connectives and defuzzification strategy. Throughout the book the first and the second topic were discussed. In this chapter focus is on defuzzification strategies. Finally, classification by IF-THEN rules with support of fuzzy queries is examined.

4.1 From Classical to Fuzzy Inference

Inference is the process of deriving logical conclusions from premises assumed to be true. Classical inference is based on the *modus ponens*. In propositional logic, *modus ponens* or implication elimination is a valid, simple argument form and rule of inference [16]. *Modus ponens* is of the structure:

$$
\begin{aligned}
&p; \\
&p \Rightarrow q; \\
&q
\end{aligned}
\tag{4.1}
$$

where p is a known fact, $p \Rightarrow q$ is a rule of inference and q is the consequence or conclusion.

For example, on Wednesdays a course of fuzzy logic takes place (IF Wednesday THEN course of fuzzy logic, $p \Rightarrow q$). Today is Monday (p), so the course is not today (q). In classical inference truth values of p, $p \Rightarrow q$ and q are either true (with logical value of 1), or false (with logical value of 0). In the truth table of implication (Table 4.1) the *modus pones* is represented by the fourth row.

© Springer International Publishing Switzerland 2016
M. Hudec, *Fuzziness in Information Systems*,
DOI 10.1007/978-3-319-42518-4_4

Table 4.1 Truth table of the classical implication

p	q	p ⇒ q
0	0	1
0	1	1
1	0	0
1	1	1

The basic rule of inference in classical logic is expressed by *modus ponens* (4.1) as [15]:

$$
\begin{array}{l}
\text{fact: x is A} \\
\text{rule: IF (x is A) THEN (y is B)} \\
\hline
\text{consequence y is B}
\end{array}
\tag{4.2}
$$

Example 4.1 A company has a rule: *IF turnover(x) ≥ 1000 (A) THEN provide a 10%(y) discount (B)* The inference is expressed as:

$$
\begin{array}{l}
\text{fact: x is turnover} \\
\text{rule: IF}(x \geq 1000)\ \text{THEN}\ (y = 10\%) \\
\hline
\text{consequence: y is discount}
\end{array}
$$

Let us have two customers C_1 with turnover of 1 000.95 and C_2 with turnover of 999.35. Obviously, C_1 meets the fact, activates the rule and receives the discount of 10 % (inference chain). Contrary, C_2 does not meet the fact and receives no discount. The inference rule is simple and clear, but the problem of two-valued evaluation of predicates appeared. The value of turnover is without doubt crisp. So, in the extreme situation company might say to the complaining customer C_2: *you know the rule, so if you had bought a small box of chewing gum in addition, you would have received discount*. Albeit not a very efficient way for keeping customers, it meets the two-valued inference.

To make inference softer, several small intervals like [0, 900)—no discount; [950, 975)—3 % of discount; [975, 1 000)—6 % of discount; [1 000, 1 050]—8 % of discount and more than 1 050—full discount of 10 % can be employed. The second option is expressing discount as a function of turnover. In order to find proper functions with suitable properties mathematical knowledge is required. □

Let us now look at a task bearing vague data. For instance, if the financial reimbursement for significantly flooded buildings should be provided by government, then a possible rule is: *IF flooded level ≥ 100 cm THEN provide a reimbursement*. Not only the sharp condition, but also the crisp value of known fact is disputable. The rough surface of the stream of moving water caused different shades of wetness on walls (Fig. 5.2). The question is, where exactly to measure. Thus, not only the sharp rule, but also crisp fact are disputable. Solving this problem by mathematical functions becomes a more complex task requiring deeper knowledge in comparison with the task from the previously mentioned example.

In inferring and many other tasks we are instantaneously faced with a continuous transition between full membership and no membership [11]. People have capability to make conclusions based on the data and rules which are often vague. It means that conclusions are made without precise measurements and calculations [52]. It means that people do not require rigid rules.

Remarking difference between crisp characteristics and variety of membership functions for the same concept (e.g. *non expensive and close to city centre hotel for student and rich businessman*). Bezdek [4] concluded

> uniqueness is sacrificed (and mathematicians howl), but flexibility is increased (and engineers smile)

On the other hand, mathematicians have significantly contributed to the formulations of fuzzy inference in order to efficiently simulate human reasoning in solving variety of tasks.

4.2   Fuzzy Inference

Inference in a fuzzy environment is realized by the generalized *modus ponens* (GMP) [6, 26, 34]. Fuzzy reasoning is an inference process that derives conclusion from a set of flexible IF-THEN rules and known facts which can be either crisp or fuzzy. In usual human reasoning, GMP is expressed in an approximate rather than crisp manner [15]:

Premise 1: x is A'

Premise 2 (rule): IF x is A THEN y is B

$$\text{(4.3)}$$

Consequence: y is B'

where A' is a fuzzy set of known fact, which is more or less equal to A in rule.

Throughout this book the word *rule* is used instead of *Premise 2* and *(known) fact* instead of *Premise 1*. The GMP differs from *modus ponens* in propositions' truth values, which are from the [0, 1] interval instead of 0 and 1. The observation *x is A'* may be fully or partially compatible with the assumption *x is A*. It causes full or partial satisfaction of the consequence *y is B'*.

4.2.1   Inference Process

A fuzzy implication in rule (4.3) is expressed as a binary relation:

$$R = A \to B \tag{4.4}$$

Thus, the inference is expressed by means of compositional rule of inference (1.45) [53]:

$$B' = A' \circ R = A' \circ (A \to B) \tag{4.5}$$

This equation is considered as the backbone for fuzzy inference systems. The compositional rule (4.5) is expressed in general case by sup-t composition [15]:

$$\mu_{B'}(y) = \sup_{x \in X} t[\mu_{A'}(x), \mu_R(x, y)], \quad \forall y \in Y \tag{4.6}$$

which is a more generalized composition than the max-min composition (1.45) discussed in Sect. 1.2.6.

For each left-continuous t-norm t and residual implication I_t (1.66) holds

$$\mu_{B'}(y) = \sup_{x \in X} t[\mu_{A'}(x), I_t(x, y)], \quad \forall y \in Y \tag{4.7}$$

All basic t-norms, except drastic product (1.50), are left continuous. Further, let hold for $(A \to B)$:

$$\mu_{A \to B}(x, y) = f(\mu_A(x), \mu_B(y)), \quad \forall y \in Y \tag{4.8}$$

where

(i) $f = t^*$, where t^* is a left-continuous t-norm or
(ii) $f = I_{t^*}$, where t^* is a left-continuous t-norm with property $t^* \geq t$ ($I_{t^*} \leq I_t$)

then compositional rule of inference (4.6) replaces GMP.

Applying residual implication and left continuous minimum t-norm t_m (1.47), we get:

$$\mu_{B'}(y) = \sup_{x \in X} \min[\mu_{A'}(x), \min(\mu_A(x), \mu_B(y))], \quad \forall y \in Y \tag{4.9}$$

which is called Mamdani method of inference shown in Fig. 4.1.

The part $\min(\mu_A(x), \mu_B(y))$ in (4.9) is often referred to as Mamdani implication, which is theoretically wrong, because minimum function does not meet all axioms of implication. But, in practice it is usual to describe implication by t-norms [18]. This especially holds for the minimum t-norm, which is often called Mamdani implication. When we are in the inference by the GMP (focused on a forth row in Table 4.1), Mamdani implication is correct. However, if we do not keep in mind that the Mamdani implication is not a full implication, then it might cause wrong results in other fields, where fuzzy implication is required. This statement was proven in managing preferences among atomic query conditions by implication in Sect. 2.4.2.

Applying residual implication and left continuous product t-norm t_p (1.48) we get:

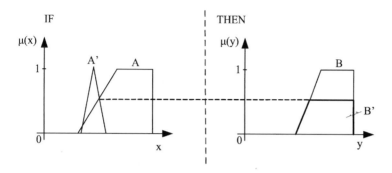

Fig. 4.1 Mamdani method of inference for single rule with single antecedent

$$\mu_{B'}(y) = \sup_{x \in X} \min[\mu_{A'}(x), \mu_A(x) \cdot \mu_B(y)], \quad \forall y \in Y \qquad (4.10)$$

which is called Larsen method of inference plotted in Fig. 4.2.

The same discussion related to the Mamdani method of inference and Mamdani implication holds for the Larsen method of inference and Larsen implication (product t-norm).

Other suitable combinations of implications and t-norms in compositional rule of inference are discussed in [7].

Example 4.2 Government has decided to reimburse villages, where high pollution was recorded during manufacturing relevant parts for a highway. Hence, the inference rule is: *IF pollution is high, THEN reimbursement is high.* The known fact is measured pollution. Inference process is graphically shown in Fig. 4.3 for the measured pollutant about 25 mg for a particular village. The fact is more or less inside

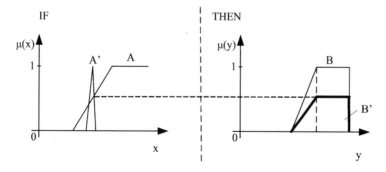

Fig. 4.2 Larsen method of inference for single rule with single antecedent

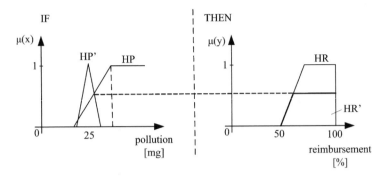

Fig. 4.3 Solution calculated by Mamdani method of inference

the concept *high pollution*, what implies that the village receives more or less high reimbursement. Village with recorded pollution of 24 mg will receive slightly smaller financial support.

The rule is local, i.e. it says nothing about managing reimbursement for small pollution. Real inference tasks usually consist of several facts and rules, therefore the formal (mathematical) procedure for inference is indispensable.   ☐

The next step is converting resulting fuzzy set B' into crisp value, because villages should receive exact amount of resources. In above-mentioned example fuzzy set A in the IF part of rule is not a fuzzy number, because support (1.9) is unlimited. It is acceptable, because government is not sure, which maximal value of pollution will be recorded. Concerning the THEN part, fuzzy numbers should be used due to two practical reasons: (i) reimbursement, discount, etc. cannot be unlimited; (ii) for the defuzzification (explained later) fuzzy set should be bounded.

In the next example, comparison of Mamdani and Larsen inference methods is provided.

Example 4.3 A company decided to improve motivation of customers, who frequently use its products. The rule is *IF frequency is high, THEN discount is high*. Inference process is graphically shown in Fig. 4.4 for the estimated frequency of 173 days of using the product by customer x. The estimated fact is more or less inside the concept of high frequency. Customer receives more or less high discount.

In the theory of t-norms holds $t_p \leq t_m$. From the properties of residual implications it implies that $I_{t_p} \geq I_{t_m}$. This fact is reflected in the inference methods. The solution obtained by Larsen method is higher than solution by Mamdani method or equal to it, depending on defuzzification methods, which are explained in the next section. A short glimpse at resulting fuzzy sets shows that HD'_M has longer flat segment than HD'_L.   ☐

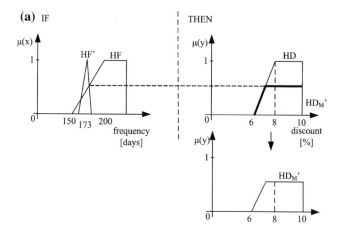

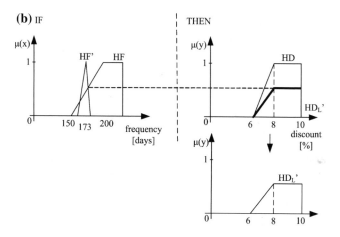

Fig. 4.4 Comparison of Mamdani and Larsen inference methods. **a** Inference procedure based on Mamdani method. **b** Inference procedure based on Larsen method

When we apply Mamdani method of inference on the single input-single rule inference (4.3), the inference procedure can be expressed as:

$$\mu_{B'}(y) = \sup_{x \in X} \min[\mu_{A'}(x), \min(\mu_A(x), \mu_B(y))] = \min[\delta, \mu_B(y)] \qquad (4.11)$$

where $\delta = \sup_{x \in X} \min[\mu_{A'}(x), \mu_A(x)]$ denotes the degree of compatibility between fuzzy sets A and A' or possibility measure (1.28) that fuzzy value A' is in concept A. The parameter δ can be also interpreted as the firing degree of a rule. This value is propagated by the rule and the resulting membership function B' cannot have greater height than δ. Thus, by the formal procedure we reached the same conclusion as by heuristic way of experience.

The further considerations are restricted to the Mamdani method of inference. The inference procedure (4.3), (4.11) can be easily generalized to the multiple inputs-multiple rules inference, which has the following structure:

Premise x_1 is A_1' AND...AND x_n is A_n'
Rules IF x_1 is A_1^1 AND...AND x_n is A_n^1 THEN y is B^1
...

IF x_1 is A_1^r AND...AND x_n is A_n^r THEN y is B^r (4.12)
...

IF x_1 is A_1^R AND...AND x_n is A_n^R THEN y is B^R

Consequence y is B'

where A_i^r ($i = 1, .., n, r = 1, ..., R$) describes linguistic term for ith attribute in rth rule and B^r stands for linguistic term expressing rth consequence.

The connectives AND are interpreted as fuzzy intersection (1.30). The aggregation of multiple rules is realized by fuzzy union (1.31). Now, applying minimum t-norm (1.47) for AND connective, Mamdani inference method (4.9) and maximum s-norm (1.59) for disjunction of rules, the following inference procedure yields:

$$\mu_{B'}(y) = \max_{r=1,...,R} \mu_B^{'r}(y) = \max_{r=1,...,R} \min[\delta_{min,r}, \mu_B^r(y)] \qquad (4.13)$$

where $\delta_{min,r} = \min(\delta_{1,r}, ..., \delta_{n,r})$ (firing degree of r-th rule) and $\delta_{i,r} = \sup_{x_i \in X_i} \min[\mu_{A_i'}(x_i), \mu_{A_i^r}(x_i)]$ denotes the degree of compatibility between fuzzy sets $A_i^{'r}$ and A_i^r.

The membership function of B' is obtained by taking the maximum of membership functions of each $B^{'r}$ clipped by the corresponding firing degree $\delta_{min,r}$. The graphic interpretation is shown in Fig. 4.5 for the case of two fuzzy facts and two rules.

If facts are measured as crisp values $(x_i', ..., x_n')$, instead of fuzzy sets $(A_1', ..., A_n')$, then procedure is somewhat simpler. The firing degree (δ) from formula for one input-one rule (4.11) becomes:

$$\delta = \sup_{x \in X} \min[\mu_{A'}(x), \mu_A(x)]$$
$$= \begin{cases} \sup_{x \in X} \min[1, \mu_A(x)], & x = x' \\ \sup_{x \in X} \min[0, \mu_A(x)], & x \neq x' \end{cases} \qquad (4.14)$$
$$= \mu_A(x')$$

Straightforwardly, formula (4.13) yields:

$$\mu_{B'}(y) = \max_{r=1,...,R} \mu_B^{'r}(y) = \max_{r=1,...,R} \min[\min_{i=1,...,n}(\mu_{A_i^r}(x_i')), \mu_B^r(y)] \qquad (4.15)$$

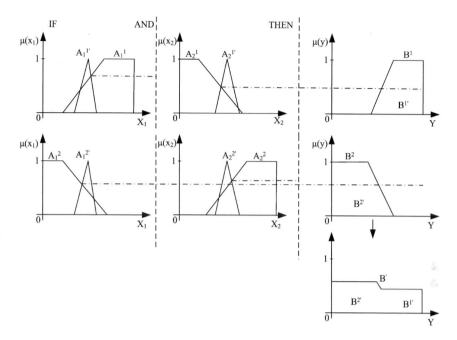

Fig. 4.5 An inference procedure for two rules and two inputs expressed as fuzzy facts

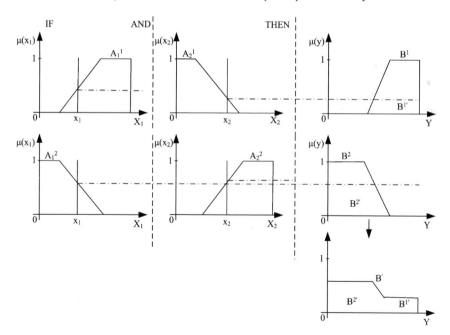

Fig. 4.6 An inference procedure for two rules and two input crisp facts

Table 4.2 Decision table of two input attributes consisted of three linguistic terms each

A_1/A_2	Small	Medium	High
Small	B^1	B^2	B^3
Medium	B^2	B^4	B^6
High	B^5	B^6	B^7

In this way, membership degrees of input values to respective fuzzy sets and selected function for the AND operator fire the rule. The inference for crisp inputs is shown in Fig. 4.6. Strictly speaking, crisp facts are expressed by fuzzy singletons, i.e. height equal to 1 (Fig. 1.10), in order to keep above discussed equations consistent.

When the rule base contains two input attributes, it can be represented by the decision table shown in Table 4.2, where B^r may be terms such as *very low*, *low*, *medium*, etc. It is worth noting that several rules may have the same output B^r. In this case the number of rules can be reduced by method shown later on.

4.2.2 Defuzzification

The consequence of inference is represented as a fuzzy set (e.g. Figs. 4.4 and 4.6) regardless of the fact, whether antecedents are fuzzy or sharp values expressed as singletons. It means that the crisp value, which in the best way represents the fuzzy set B' must be calculated. In business system values, such as bonus, discount or reimbursement should be crisp values. In a technical system a valve should be turned by exact angle.

Fuzzy conjunction can be expressed by many t-norm functions, fuzzy implication can be strong, quantum, residual, each expressed by variety of functions. The same holds for defuzzification. In this book only the best known methods are examined. Defuzzification strategies are deeply discussed, e.g. in [13, 29, 40, 41, 47, 51].

Commonly used strategies for calculating crisp value $y^0 \in Y$ that represents $B' \in F(Y)$, where $F(Y)$ is a family of fuzzy set on the domain of attribute appearing in consequence, are [28]:

- **Maximum value strategies** or extreme value strategies considering only flat segment or point(s) of $\mu_{B'}(y)$ with the maximal value of height (1.11)
- **Gravity strategies** considering the entire shape of the membership function $\mu_{B'}(y)$

Maximum (extreme) value strategies

If the $\mu_{B'}(y)$ is unimodal (i.e. has unique maximum), the solution is simple: y^0 is equal to height of $\mu_{B'}(y)$ [55], i.e.:

$$y^0_{unm} = \sup_{y \in Y} \mu_{B'}(y) \qquad (4.16)$$

For non-unimodal $\mu_{B'}(y)$, several methods, which calculate the defuzzification value on the flat segment of the maximal height expressed as:

$$F_s(y) = \{y | y \in Y \wedge \neg(\exists z \in Y)(\mu_{B'}(z) > \mu_{B'}(y))\} \qquad (4.17)$$

are [55]:

- Left of Maxima (LOM)—where the defuzzified value y_0 gets value

$$y^0_{lom} = \min\{y | y \in F_s\} \qquad (4.18)$$

- Right of Maxima (ROM)—where the defuzzified value y_0 gets value

$$y^0_{rom} = \max\{y | y \in F_s\} \qquad (4.19)$$

- Center of Maxima (COM)—where the defuzzified value y_0 gets value

$$y^0_{com} = \frac{\min\{y | y \in F_s\} + \max\{y | y \in F_s\}}{2} \qquad (4.20)$$

The LOM, ROM and COM strategies are illustrated in Fig. 4.7. Clearly, $y^0_{lom} \leq y^0_{com} \leq y^0_{rom}$. The COM method should not be confused with the Mean of Maxima (MOM) method, which assumes that there is not a flat segment F_s (4.17), but separated different maxima [55].

Example 4.4 Inference by Mamdani and Larsen methods are illustrated in Example 4.3. Crisp values that represent output fuzzy set HD' for both methods are:
$$y^0_{lom-M} = 7 < y^0_{lom-L} = 8$$
$$y^0_{com-M} = 8.5 < y^0_{com-L} = 9$$
$$y^0_{rom-M} = 10 = y^0_{rom-L} = 10$$

The result is not surprising, because the flat segment calculated by Mamdani method starts in smaller value of y (caused by minimum t-norm) than the flat segment of Larsen method, which is caused by product t-norm. Both flat segments depicted in Fig. 4.4 end in the same value of Y. □

Fig. 4.7 Comparison of maximum (extreme) values defuzzification strategies

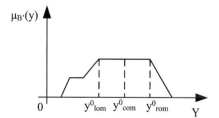

The **gravity strategies** are represented by:

- Center of Gravity (COG) calculated as:

$$y_{cog}^0 = \frac{\int_y y \cdot \mu_{B'}(y)}{\int_y \mu_{B'}(y)} \tag{4.21}$$

is widely adopted defuzzification strategy. Formula (4.21) is Rieman integral. In usual tasks of inference, no problems exist as for solving these integrals.
- Half of Field (HOF), y_{hof}^0 is the solution of

$$\int_{-\infty}^{y_{hof}} \mu_{B'}(y) = \int_{y_{hof}}^{\infty} \mu_{B'}(y) \tag{4.22}$$

i.e. y_{hof}^0 is a vertical line, which partitions the field under $\mu_{B'}(y)$ in half.

Maximum values strategies are less demanding for calculation, but less precise, because they do not consider areas of lower membership degrees. Hence, they are especially suitable for "rough" inference, where the solution is reached without complex calculations of integrals. The gravity strategies are suitable for tasks, where the sensitivity of defuzzified value is crucial. If flat segment is more or less symmetrical to the whole shape of $\mu_{B'}(y)$, then $y_{com}^0 \approx y_{cog}^0$ (Fig. 4.8—upper graph) holds.

Otherwise, values of y_{com}^0 and y_{cog}^0 can significantly vary (Fig. 4.8—lower graph). A mixture of both strategies is

- Height Defuzzification Method (HDM), which is a generalization of COM method [8]. It uses all flat segments of $\mu_{B'}(y)$. Thus, defuzzified value y_0 is:

Fig. 4.8 Comparison of COG and COM

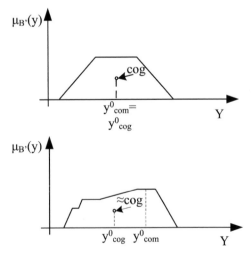

$$y^0_{hdm} = \frac{\sum_{i=1}^{n} h_i \dfrac{y_{1i} + y_{2i}}{2}}{\sum_{i=1}^{n} h_i} \qquad (4.23)$$

where h_i is the height of ith flat segment $[y_{1i}, y_{2i}]$. For non-symmetric output fuzzy sets this measure is closer to COG (4.21), than other Maximum values strategies, because operation of defuzzification is performed not only on flat segments of maximal height.

Example 4.5 The fuzzy inference rule for managing reimbursement for highly polluted villages has been created in Example 4.2 (Fig. 4.3). Let us the set *high pollution* is expressed as R fuzzy set (Fig. 1.9) with parameters $a = 90$ mg and $m_1 = 100$ mg. The government has decided that *high reimbursement* is defined by parameters $a = 600$, $m = 800$ and $b = 1000$. The measured pollution for four municipalities is $M1$: 85 mg, $M2$: 95 mg, $M3$: 97.5 mg and $M4$: 108 mg. For the inference the Mamdani method (4.9) is chosen. Concerning $M1$, firing level is 0 and therefore rule is not activated. The inference for $M2$ is shown in Fig. 4.9. Thus, by (4.11) we got $\mu_{B'}(y) = \min[0.5, \mu_B(y)]$. Set $\mu_{B'}(y)$ has core in the interval $[700, 1000]$. Hence, by the COM defuzzification method (4.20) the reimbursement is $y^0 = \frac{700+1000}{2} = 850$. In the same way, reimbursement of $M3$ is 875 and for $M4$ is 900. When the Larsen method (4.10) of inference is chosen, then the reimbursement is higher, due to the shorter flat segment.

Let us look at $M2$ and the three methods of the maximal value strategy. If the ROM (4.19) is chosen, the municipality will receive reimbursement of 1000. But, if the LOM (4.18) is chosen, the municipality will receive reimbursement of 700. The value

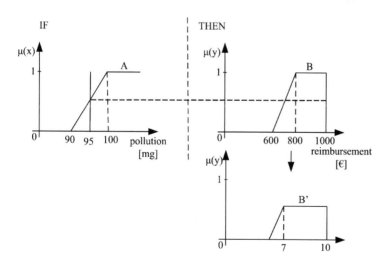

Fig. 4.9 Reimbursement for a highly polluted municipality $M2$

of COM is somewhere in middle causing that both sides could be satisfied. By the ROM strategy every municipality which activates the rule will receive reimbursement of 1000. This holds for the one input one rule case.

The inference rule might remain stable during a long period. The linguistic interpretation of rule remains clear and reasonable. Only the parameters of fuzzy sets should be adjusted to the new knowledge about limits of particular pollutants related to their impact on health.   □

Defuzzification strategies may be realized in two ways [3, 12]: Mode A-FATI (first aggregate then infer) and Mode B-FITA (first infer then aggregate). In the former, defuzzification procedure consists of two steps: (i) aggregation of individual $B^{r'}$ into an overall fuzzy set B' by, e.g. (4.13); (ii) one of aforementioned defuzzification strategies D is applied on B', i.e. $y^0 = D(\mu_{B'}(y))$. This way is illustrated in Figs. 4.5, 4.7 and 4.8.

In the latter, the contribution of each $B^{r'}$ is defuzzified separately and then y^0 is computed as weighted average:

$$y^0 = \frac{\sum_{r=1}^{R} \delta^r y^r}{\sum_{r=1}^{R} \delta^r} \tag{4.24}$$

where y^r is defuzzified value of fuzzy set $B^{r'}$ and δ^r is the matching degree between the input observation and the premise of rule r.

The Mode B-FITA way causes better response time due to reduced computational effort [31].

4.2.3 Illustrative Examples and Issues

The next examples illustrate an inference task consisting of two input variables and four rules.

Example 4.6 An institute faces a task of estimating needs for improvement of environmental situation. Due to uncertainties in rules, a fuzzy inference system is considered. Two main input attributes are *population density* and *pollution of selected pollutant*. For illustrative purpose input attributes are fuzzified into two fuzzy sets: *small* and *high*. Output variable: *needs for resources* is fuzzified into three fuzzy sets: *small, medium* and *high*. Fuzzy sets of input variables and consequence are plotted in Fig. 4.10.

The number of rule is four and the number of possible outputs is three. Therefore, the inference rules can be presented in the structure of IF-THEN rules:

- IF population density is small AND pollution is small THEN reimbursement is small
- IF population density is small AND pollution is high THEN reimbursement is medium

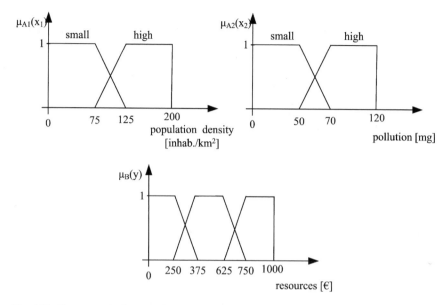

Fig. 4.10 Fuzzy sets for input and output attributes

- IF population density is high AND pollution is small THEN reimbursement is medium
- IF population density is high AND pollution is high THEN reimbursement is high

or in the decision table shown in Table 4.3, where rows represent population density, columns represent pollution and inner cells represent outputs.

The rule evaluation is realized by Mamdani method of inference (4.9). The needs for municipality M with population density of 145 inhabitants/km^2 and pollution of 65 mg is plotted in Fig. 4.11. The municipality fully belongs to the set high population density and partially to sets small and high pollution. Hence, two rules are fired with $\delta < 1$. The set B' consists of cuted sets *medium* and *high*. The final step is defuzzification. Results yielded by several defuzzification strategies are in Table 4.4. For the municipality, which activates only the fourth rule, the result by all strategies except ROM will be greater than for municipality M. Hence, we can say that from municipality perspective ROM is optimistic, LOM pessimistic and COM somewhere in between. For the agency providing resources the perspective is just the opposite: ROM is pessimistic and LOM optimistic. Hence, compromising results are obtained by COM and HDM.

Table 4.3 IF..AND..THEN rules for improving ecological situation

Pollut./Pop. dens.	Small	High
Small	Small	Medium
High	Medium	High

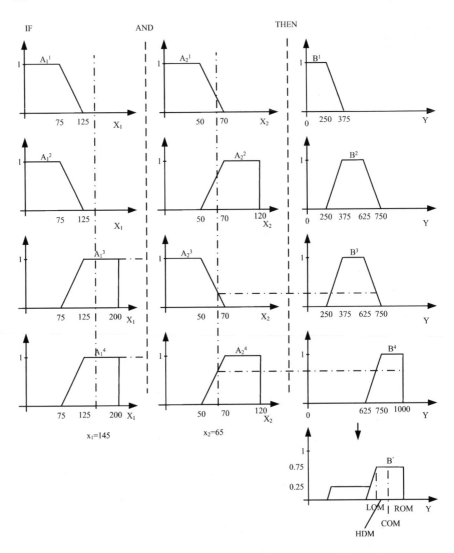

Fig. 4.11 An inference based on two attributes, three output fuzzy sets and four rules

Needs for other municipalities are calculated in the same way. It ensures that similar municipalities receive similar resources. If no rule is activated, it means that either tuple is outlier, or the quality of rule base is problematic.

It should be emphasized that municipality with population density of 100 inhabitants /km^2 and pollution of 60 mg will activate all four rules with level of 0.5. Hence, output is a flat segment on the whole domain having height of 0.5. All methods which deal with averages (COM, HDM, COG) get the same result in the middle of domain having a consequence that this municipality receives support of 500, which

Table 4.4 Needs obtained by different defuzzification strategies for municipality M

Defuzz. Strategy	y^0
ROM (4.19)	1000
COM (4.20)	859.37
HDM* (4.23)	761.72
LOM (4.18)	718.75

$$*HDM = \frac{0.75 \cdot \dfrac{1000 + 718.75}{2} + 0.25 \cdot \dfrac{656.25 + 281.25}{2}}{0.75 + 0.25} = 761.72$$

is fully acceptable solution. But, LOM suggests zero support, whereas ROM the full support.

Users have the freedom to select defuzzification strategy, but should be able to defend the decision. Roughly speaking, if left part of consequence space is activated, LOM suggests no support; if right part of consequence space is activated, ROM suggests full support. If it is an acceptable solution, then LOM and ROM are options. Otherwise gravity strategies like COM and HDM should be options to consider. □

It is no surprise that the question, which defuzzification strategy is the most suitable, has appeared. It is because we were faced several times with the same question (which t-norm, which s-norm, which implication, etc.). Generally, gravity strategies are suitable, especially COG. Sophisticated tools offer these strategies, among others. If company wishes to create its own less demanding tool, then procedures calculating extreme values strategies can be considered. Furthermore, it is on company's decision, how much resources is willing to provide, e.g. for motivating customers, and therefore LOM and ROM could be considered as well.

Output variables can be expressed as any type of fuzzy set with bounded support. It implies that singletons (Fig. 1.10) are also acceptable. The inference process remains unchanged. The difference is in simplified defuzzification procedure illustrated in Example 4.7.

Example 4.7 A company is going to motivate customers by internal coupons with different values according to bought items (price) and payment delay. Company has created the following rule base:

- IF turnover is small AND payment delay is high THEN coupon is small
- IF turnover is small AND payment delay is small THEN coupon is medium
- IF turnover is high AND payment delay is high THEN coupon is medium
- IF turnover is high AND payment delay is small THEN coupon is high

Assume that instead of usual three fuzzy sets *small*, *medium* and *high*, company has decided to use singletons to model the output discount as:

- IF turnover is small AND payment delay is high THEN coupon is 0^*
- IF turnover is small AND payment delay is small THEN coupon is 5

Table 4.5 Customers with turnover and payment delay

Customer	Turnover [€]	Payment delay (days)
c1	740	13
c2	1000	10
c3	1700	9
c4	1350	7

- IF turnover is high AND payment delay is high THEN coupon is 5
- IF turnover is high AND payment delay is small THEN coupon is 10

* numbers are used instead of singletons' names.

Let us state for the sake of simplicity, four customers with the values of turnover and payment delay exist in the database run by company (Table 4.5).

The inference model for customer $c3$ is shown in Fig. 4.12. Two rules are fired affecting two singletons (heights of $\mu_B^3(y)$ and $\mu_B^4(y)$). The consequence is unimodal non-convex fuzzy set. The simplest method of defuzzification (4.16) picks value of 10 as a solution. It means that this customer receives coupon of value of 10. Though, this customer is a perspective one for the company, there is another one ($c4$) with better performances, who receives the same value of coupon. Let us further speculate that customers $c3$ and $c4$ know each other. Customer $c4$ could consider this way of motivation as unfair and decides to buy products from the competitor.

The better option is defuzzification by HDM (4.23), where all activated singletons affect the solution. The solution is in Table 4.6.

The motivation is better tailored to customers (more fairly distributed), the meaning of the rule base is clear at the first glance and singletons on output ensure simpler calculations.

If one customer has significantly much higher turnover than the others, then (with other part of rule satisfied as well) it should receive maximal reward. For this reason, fuzzy set *high* in the antecedent part should not be limited from the right side.   □

The difference between crisp value and fuzzy singleton is in vertical representation (membership degree). Singleton is a fuzzy set and therefore different firing level of considered rules causes different height of fuzzy singletons in output, as was illustrated in the aforementioned Example 4.7.

The critical part is the construction of fuzzy sets for input and output attributes. It is desirable that the family of fuzzy sets covers the whole domain of considered attributes in order to catch all possible observations. In illustrative examples in this section two fuzzy sets cover the whole domain of attributes. In real-life tasks, usually three to seven fuzzy sets cover the domains. Fuzzy sets can be created by domain expert(s), designers of fuzzy inference system or mined from the data by variety of approaches, such as linguistic summaries (Sect. 3.7.2), neural networks, genetic algorithms and the like.

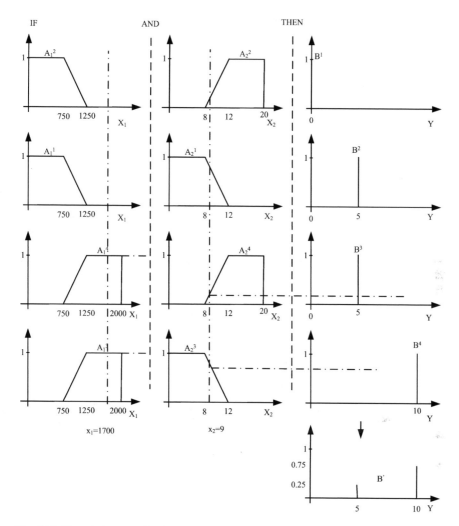

Fig. 4.12 Fuzzy inference with singleton outputs—case of customers motivation

Table 4.6 Coupons adjusted to customers by (4.23)

Customer	Coupon [€]
c1	0
c2	5
c3	8.75
c4	10

Theoretically, practitioners or domain experts are not required to know mathematical background of the inference process. Many software products for inference exist. But, on the other hand, the basic knowledge is welcome, because then they can use the complex tools more intuitively and can be aware of differences in existing functions for conjunction, disjunction, implication and defuzzification.

4.3 Fuzzy Inference Systems

These systems are computing frameworks, which reach the conclusion by concepts of fuzzy set theory, fuzzy IF-THEN rules and inference processes described in Sect. 4.2. These systems are known by other names, such as: fuzzy logic controllers, fuzzy expert systems, fuzzy rule-based systems, fuzzy associative memories and by simplified, but rather ambiguous name: fuzzy systems [23].

While software packages focused on flexible querying or mining summaries from data support the experts in their work by data and summarized information, a fuzzy inference system is expected to model expert's knowledge and make it available (to non-experts) for decision making, diagnosis, technical systems control and the like. Expert system may be defined as:

> An expert system is a computer program that solves problems that heretofore required significant human expertise by using explicitly represented domain knowledge and computational decision procedures [25].

Fuzzy inference systems were initially developed for controlling technical systems. Pioneers in theoretical experiments and practical realization were Mamdani and Assilian [33] who realized control of a laboratory model of steam engine. The first industrial application was a control of cement kiln. Takagi, Sugeno and Kang [43, 44] have created an inference system, in which consequences are functions instead of fuzzy sets. In this section main points of these models are outlined. More about these models and their comparison can be found, e.g. in [17, 24, 54].

4.3.1 Mamdani Model (Logical Model)

In this model both inputs and consequences are expressed as fuzzy sets. This system works on principles discussed in Sects. 4.2.1 and 4.2.2. Both the inputs and consequences in each fuzzy rule are fuzzy sets, which express linguistic terms. Hence, this type is also called logical model of inference. The structure of the model envelopes four conceptual components shown in Fig. 4.13. The main part is the fuzzy inference engine, which performs the inference procedure based on the Mamdani method of inference (4.9) on known fact (crisp or fuzzy) and rule base strictly following the compositional rule of inference. Because the Mamdani model produces outputs as fuzzy sets (including singletons), the defuzzification method should be used. The

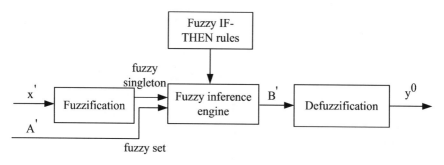

Fig. 4.13 Block diagram of the Mamdani model based on [15]

Mamdani type is represented as

$$R^r : \text{IF } x_1 \text{ is } A_1^r \text{ AND...AND } x_n \text{ is } A_n^r \text{ THEN } y^r \text{ is } B^{lr} \qquad (4.25)$$

If x is crisp value, then $A'(x)$ is computed as:

$$\mu_A'(x) = F(x_0) \qquad (4.26)$$

where F is a fuzzification operator and x_0 is crisp input. The usual choice is the singleton fuzzification, i.e.
$$\mu_A'(x) = \begin{cases} 1 & \text{for } x = x_0 \\ 0 & \text{otherwise} \end{cases}$$

Non-singleton options [39] like triangular or Gaussian fuzzy sets are suitable for representing imprecision of measurements. This imprecision is examined in Chap. 5.

Two drawbacks of rule-based systems, which may appear in practice, are [31]:

- When the input attributes are mutually dependent, it is difficult to find proper fuzzy granulation of the input space.
- The size of a rule base increases rapidly with the number of input attributes and linguistic terms constructed on their respective domains. This problem is known as the *course of dimensionality*. For instance, if a rule base consists of three input attributes fuzzified into five linguistic terms each, the total number of rules is 125. This issue complicates the interpretability of the system by user.

These issues can be mitigated by making the rule base more flexible. One of solution is DNF (disjunctive normal form) fuzzy rule base, which has the following form [32]:

$$\text{IF } x_1 \text{ is } \tilde{A}_1 \text{ AND...AND } x_n \text{ is } \tilde{A}_n \text{ THEN } y \text{ is } B \qquad (4.27)$$

where each input value x_i takes as its value a set of linguistic terms $\tilde{A}_i$, whose members are connected by a disjunctive aggregation, that is, x_i is $\tilde{A}_i = \{A_{i1} \text{ or } A_{i2} \text{ or ... or } A_{ip}\}$.

Hence, the rule of a DNF structure may be:

IF x_1 is $\{A_{11}\}$ AND x_2 is $\{A_{21}$ or A_{22} or $A_{23}\}$ AND x_3 is $\{A_{31}$ or $A_{32}\}$ THEN y is B.

For example, an observation may state: *IF length of roads is very high THEN winter road maintenance is high regardless of the value of number of days with snow coverage and precipitation.* This observation is converted into the rule of a DNF structure:
IF length of roads is high AND number of days with snow is {small OR medium OR high} AND precipitation is {small OR medium OR high} THEN maintenance is very high.

4.3.2 Sugeno Model (Functional Model)

The structure of this model envelopes two conceptual components shown in Fig. 4.14. This model is similar to the Mamdani method (4.25) in several points. The IF part and rule base are exactly the same. The difference is in the consequence. In the Sugeno model consequence is either constant, or linear function, i.e. rule structure that has fuzzy antecedent and functional consequent parts. Therefore, this model is called functional model of inference. The system is represented as:

$$R^r : \text{ IF } x_1 \text{ is } A_1^r \text{ AND...AND } x_n \text{ is } A_n^r \text{ THEN } y^r = a_1^r x_1 + ... + a_n^r x_n + c^r \quad (4.28)$$

where variables in IF part have the same meaning as in (4.12), a_i^r is ith coefficient in rth rule and c^r is a constant. In a zero-order Sugeno model, the consequence of rule i (y^i) is a constant (all coefficients a_i, $i = 1, ..., n$ gets value of 0). Otherwise, Sugeno model is the first-ordered model.

Unlike the Mamdani model, the Sugeno model cannot strictly perform compositional rule of inference [15]. Since each fuzzy rule has a crisp output, the overall output y^0 is calculated via weighted average:

$$y^0 = \frac{\delta_{min,1} \cdot y^1 + ... + \delta_{min,R} \cdot y^R}{\delta_{min,1} + ... + \delta_{min,R}} \quad (4.29)$$

Fig. 4.14 Block diagram of the Sugeno model based on [15]

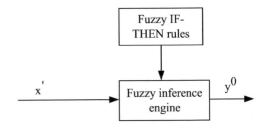

or weighted sum:

$$y^0 = \delta_{min,1} \cdot y^1 + \dots + \delta_{min,R} \cdot y^R \tag{4.30}$$

where $y^r = f^r(x_1', \dots, x_n'), r = 1, \dots, R$ and δ has the same meaning as in (4.11) and (4.13). This equation corresponds with the B-FITA defuzzification (4.24), because both are related to interpolation, where y^r is individual output of each rule either in functional way, or as defuzzified value of $B^{r'}$.

If no rule is activated, then weighted average ends in infinity. Concerning the weighted sum, result is 0. Care should be focused on constructing fuzzy sets, especially for input variables. If a tuple (e.g. customer or territorial unit) does not activate any rule, it cannot be assigned to any output fuzzy set and therefore cannot be properly motivated or reimbursed, for example. Hence, all possible input values should be considered.

An illustrative example of rules in Sugeno model is:

- IF temperature (x_1) is high AND pressure (x_2) is high THEN set valve on $2x_1 + x_2$
- IF temperature (x_1) is low AND pressure (x_2) is low THEN set valve on $3x_1 - \frac{1}{2}x_2$

Practically, a zero ordered Sugeno model is equivalent with the Mamdani model expressed by singletons on the consequence side. But, theoretically, though the singleton seems to be an ordinal number, it is expressed as fuzzy set and therefore firing degree δ can cut its height, so we should apply defuzzification method. On the other side, a zero-ordered Sugeno model uses classical numbers. The weighted average of crisp outputs in Sugeno model (4.29) is equivalent to the HDM defuzzification method (4.23) applied on singletons.

Example 4.8 Let us look again at Example 4.7. This inference can be solved by a zero-ordered Sugeno model. The rule base is:

- IF turnover is high AND payment delay is small THEN coupon is 10
- IF turnover is high AND payment delay is high THEN coupon is 5
- IF turnover is small AND payment delay is small THEN coupon is 5
- IF turnover is small AND payment delay is high THEN coupon is 0

Let us focus on customer $c2$ from the Example 4.7. This customer activates all four rules with $\delta = 0.5$. By weighed average (4.29) the solution is 5. The same as in Mamdani model with singleton outputs and HDM method. But, if weighted sum (4.30) is used then the customer $c2$ receives output of 10. □

4.4 Fuzzy Rule-Based System Design

Two components of fuzzy rule-based system are knowledge base and fuzzy inference. The latter, discussed in Sect. 4.2, is related to formal mathematical structures of inference, which are able to imitate human' reasoning. The former serves as a repository of the problem-specific knowledge upon which the interface process reaches the output for the observed inputs [31].

Roughly speaking, building the fuzzy rule-based system is divided into three main steps: deciding which model to employ (Mamdani or Sugeno), constructing the rule base and selecting defuzzification strategy.

The first step in design of a knowledge base is deciding, whether to employ Mamdani or Sugeno model. If designer wishes to avoid defuzzification and output is better explained by functions, then Sugeno model is the choice. Otherwise, Mamdani model is the right one.

The next step is related to building rule base: to decide which input attributes are related to the output one, fuzzification of their domains into linguistic terms and merging input attributes with output attribute by IF-THEN rules. Domains of selected attributes are fuzzified into several fuzzy sets. In most cases this number ranges from three to nine, with nine being an upper limit according the observations expressed in [38].

Fuzzy sets expressing attributes are often and, sometimes reasonably, designed with insufficient information about behaviour of attribute A at the beginning of the rule base design [45]. In the sketch of the rule base, membership functions represent more or less rough approximation. Attributes may be influenced by variety of facts. Designers of inference systems or experts should be aware that the more information related to input attributes A_i and the output ones is available, the more close the form of membership function $\mu_{A_i}(x)$ is to ideal representation. Hence, construction of fuzzy sets should be done carefully. In order to meet this goal Trillas and Morega [46] suggested the following steps, which may be useful in construction of a rule base:

1. Capture the rough behaviour of attribute A in regard to its universal set X
2. Once a shape of a prototype of the membership function is reached, the final form depends on more information about parameters and shape of fuzzy set. This information should be deeply searched.
3. Based on additional information obtained in Step 2, the new membership function $\mu_A(x)$ is constructed
4. This membership function should be checked against the known data to be sure that it satisfies what is required
5. In case of positive answer, $\mu_A(x)$ is accepted. If negative answer appears, designers should return to Step 1.

Commonly accepted way for overlapping these sets is in their crossover point (1.14). An example are the fuzzy sets plotted in Fig. 4.10.

Measures of fuzziness can be helpful in evaluation of constructed fuzzy sets. If this measure is equal to 0, then created sets are crisp. On the other side, maximal value of this measure (if measure of fuzziness for a set and its complement is equal) reveals that the uncertainty is unacceptable high, i.e. instead of the fuzzy sets *small*, *medium*, *high* we have one fuzzy set *medium* with the flat segment of value 0.5 over the domain. Both extremes are illustrated in Fig. 1.19.

Aforementioned aspects are illustrated on the model for rewarding good sellers.

Example 4.9 The goal is to create fair and flexible policy for rewarding sellers in a company. Hence, the output attribute is *reward* in money units. Acceptable input attributes are *number of sold items* and *average persuasion time*. Furthermore, both attributes are equally important. For the sake of simplicity, let both input attributes be fuzzified into two fuzzy sets and output variable into three fuzzy sets. Knowing these facts, the rule base may be of structure:

- IF number of sold items is high (A_1^1) AND persuasion time is short (A_2^1), THEN reward is high (B^1)
- IF number of sold items is high (A_1^2) AND persuasion time is long (A_2^2), THEN reward is medium (B^2)
- IF number of sold items is small (A_1^3) AND persuasion time is short (A_2^3), THEN reward is medium (B^3)
- IF number of sold items is small (A_1^4) AND persuasion time is long (A_2^4), THEN reward is small (B^4)

Let us consider predicate *number of sold items is high* for which we need to construct fuzzy set *high*.

This predicate is expressed by increasing function, which may be linear one with unlimited support and core (1.9), because we are not aware of the possible highest number of sold items. Fuzzy set *high reward* is also expressed by increasing function, but with the limited core and support, because the maximal value of reward cannot be arbitrary high.

If company has sellers in several regions, then external factors, such as unemployment and climatic conditions, which may affect number of sold items and persuasion time, should be taken into account. Hence, parameters of fuzzy set (a, m) in (1.23) differs by regions. Sellers in regions of low unemployment and suitable climatic conditions have better external factors. By external factors we mean attributes, the values of which are not collected in company database, but may be available from e.g. statistical office. In this way the same structure of the rule base is used in each region. The difference is in parameters of fuzzy sets, which may be stored in fuzzy relational databases and used, when sellers from a particular region are evaluated.

Having obtained this information, membership functions for input linguistic terms are constructed for each region. A suitable way for the construction is the uniform domain covering method discussed in Sect. 2.2. Membership functions for output linguistic terms can be the same for all regions. These functions can be checked by measures of fuzziness—Sect. 1.2.5 or cardinalities—Sect. 1.2.1 to see, how they cover respective domains.

The rule base states that there are two rules with the same output and both input attributes are equally important. If it is not the case, the second rule may have output *medium-high* and the output of the third rule may be *medium* stating that *number of sold items* is more relevant. □

Let the decision be Mamdani model, because linguistic interpretation of output is better legible for decision makers. The defuzzification strategy should be chosen as well. Discussion related to strategies and their properties can be found, e.g. in [55].

Short discussion in Example 4.6 may be also helpful. The COM or COG strategies are suitable, because the whole flat segment or the whole shape respectively are reflected in the final value of reward.

Computational effort may be considered as well. The decision depends on considering, whether the task requires fast method (for example, control of a technical system in real time), or time is not a crucial element (for instance, decision-making). In Example 4.9 time is not the crucial element, so the A-FITA is an option for the defuzzification mode.

4.5 Fuzzy Classification

The classifications and rankings of objects (customers, municipalities,...) are topics which gain increasing interest of decision and policy makers and researchers. We could say that:

> In the classification, objects are classified into several classes what enables better overview of all objects and a particular action could be undertaken on objects from a chosen class [19].

4.5.1 A View on Crisp Classification

Classification can be realized by IF-THEN rules. Crisp classification consists of precise values and sharp rules. Assume classification space created by four rules:

- IF $A_1 < Q$ AND $A_2 < R$ THEN x belongs to C_1
- IF $A_1 \geq Q$ AND $A_2 < R$ THEN x belongs to C_2
- IF $A_1 < Q$ AND $A_2 \geq R$ THEN x belongs to C_3
- IF $A_1 \geq Q$ AND $A_2 \geq R$ THEN x belongs to C_4

plotted in Fig. 4.15. One of drawbacks of crisp classification is the sharp distinction between similar tuples $r2$ and $r3$ and the same treatment of tuples $r1$ and $r2$, as well as tuples $r3$ and $r4$, even though they have not similar values of considered attributes. The same holds for tuples $r5$ and $r6$. This drawback is further discussed in [36].

A possible mitigation of the recognized issue can be realized by more classes plotted in Fig. 4.16. In such a classification space tuples $r1$ and $r2$ as well as $r3$ and $r4$ are now distinguishable, but the number of rules significantly increased. The new rule base consists of 16 output classes expressed by sharp boundaries.

On the other side, when rules have to be crisp without any doubt (e.g. classify customers, whether they met or did not meet the deadline for payment), then crisp classification efficiently solves such a task.

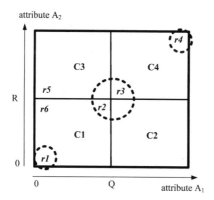

Fig. 4.15 Classification space of crisp classes

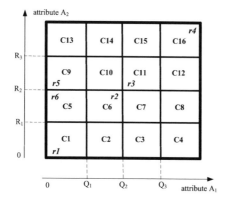

Fig. 4.16 Relaxed crisp classification space from Fig. 4.15 with more rules

4.5.2 Managing Fuzzy Classification

Fuzzy classification systems are a counterpart of above discussed fuzzy control systems. The IF part of classifiers is expressed by fuzzy sets. Output takes values from a finite set of possible values representing classes. When object must belong to one class, the winner criteria rule may be applied. The overall output is assigned to the consequence of rule having the highest firing value. This way is suitable for instance, in voting systems or in assigning suitable position for workers in a company. Concerning reward in Example 4.9, better option are overlapped classes.

The problem of crisp classification illustrated in Fig. 4.15 can be solved by overlapped output classes shown in Fig. 4.17 [36]. Now, tuples $r2$ and $r3$ belong to all four classes with different membership degrees. Furthermore, tuples $r1$ and $r2$ as well as $r3$ and $r4$ are distinguishable. Instead of large number of crisp rules, we have four fuzzy IF-THEN rules:

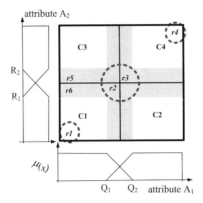

Fig. 4.17 Relaxed crisp classification space from Fig. 4.15 by overlapping classes

- IF A_1 is small AND A_2 is small THEN x belongs to C_1
- IF A_1 is high AND A_2 is small THEN x belongs to C_2
- IF A_1 is small AND A_2 is high THEN x belongs to C_3
- IF A_1 is high AND A_2 is high THEN x belongs to C_4

where output classes can be fuzzy sets, fuzzy singletons or constants.

Fuzzy classification also copes with drawbacks [37]. The definition of fuzzy sets for input attributes and output classes remains to be a challenging task. The design of fuzzy classes requires the experts or users to be aware of this issue.

If a rule base inappropriately covers data, then its quality is low. Let us examine this issue on a classification space, presented in Fig. 4.18, which consists of two input attributes and four output classes. Domains for both attributes are divided into two fuzzy sets *small* and *high* (Fig. 4.18 upper graph) taking into account the whole domains defined in the database. Fuzzy sets are constructed applying the concept of building linguistic variables (Sect. 1.4) and fuzzy partitions (1.70). Let for both attributes data distribution $f_A(x)$ be similar like presented in Fig. 4.18 on graph in the middle. As a consequence, majority of records are in class $C1$, while class $C4$ is almost empty (Fig. 4.18 bottom graph). In this way, majority of entities is treated (e.g. motivated) in the same way. It means that the classification task is an unnecessary burden, because this classification produces no significant effect, i.e. the main goal of classification is not met.

The quality measures proposed in Sect. 3.6.1, especially validity expressed by proportion, coverage (3.12) and outliers (3.14) are able to detect this problem. Based on this information, user can adjust parameters of fuzzy sets or modify number of rules.

Example 4.10 The task is classifying municipalities with regard to the estimation of the winter road maintenance needs using attributes length of roads (Road) and number of days with snow coverage (Snow). The domain of attribute number of days

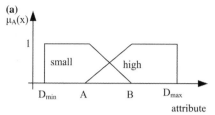

Family of fuzzy sets on domain of attribute

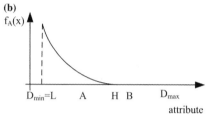

Distribution of stored data in database

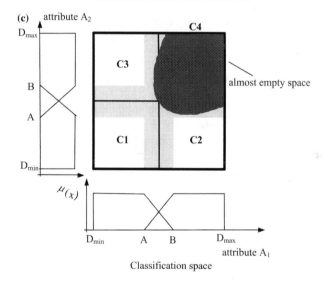

Classification space

Fig. 4.18 Inappropriately defined fuzzy sets

with snow is the interval [0, 365] of integers. The domain for the length of roads attribute is a set of real numbers greater than zero.

The rule base for the experiment contains nine rules of the IF-THEN structure:

- IF Road is Small AND Snow is Small THEN Needs is C^1
- IF Road is Small AND Snow is Medium THEN Needs is C^2
- IF Road is Small AND Snow is High THEN Needs is C^3
- IF Road is Medium AND Snow is Small THEN Needs is C^4
- IF Road is Medium AND Snow is Medium THEN Needs is C^5

Table 4.7 Distribution of municipalities over classes

Output class	Number of municipalities fully belonging	Number of municipalities partially belonging
C^1	253	157
C^2	131	167
C^3	13	19
C^4	3	18
C^5	2	11
C^6	0	2
C^7	0	4
C^8	0	4
C^9	0	0

- IF Road is Medium AND Snow is High THEN Needs is C^6
- IF Road is High and AND is Small THEN Needs is C^7
- IF Road is High and AND is Medium THEN Needs is C^8
- IF Road is High and AND is High THEN Needs is C^9

The percentage of needs for the winter road maintenance can be associated with each fuzzy output class: for instance, the class C^1 gets a percentage of needs of 10 %, C^2 gets 20 % ... and C^9 gets 90 %. For example, the distribution of municipalities over classes $C^1 - C^9$ is illustrated in Table 4.7. It means that the majority of municipalities is situated in classes C^1 and C^2, while no single municipality fully belongs to classes $C^6 - C^9$. It could mean that fuzzy sets, classes or rules are not appropriately defined. Clearly, the rule base is of low quality and should be modified.

□

The rule base from Example 4.10 can be modified in several ways. One of possibilities is creating a rule base containing more attributes and rules, but less output classes in the structure of DNF rule base.

The topic of fuzzy sets construction related to rule-based systems is covered by vast literature, e.g. [5, 14, 21, 22, 42, 48, 50].

4.5.3 Fuzzy Classification by Fuzzy Queries

Roughly speaking, data selection is a special case of classification, where tuples are separated into two classes: relevant ones that are selected and irrelevant that are not selected. It is well known that non-weighted fuzzy rules can be directly translated into fuzzy queries [9]. Since queries' conditions are equivalent with the IF part of rules and results are tuples that fully or partially belong to the output classes, the classification query language can be designed in the spirit of the fuzzy querying approaches described in Chap. 2. Hence, the results of queries are tuples selected into overlapping output classes. Finally, an aggregation should be applied. Fuzzy classification by fuzzy queries is examined, e.g. in [10, 19, 36, 49].

The fuzzy Classification Query Language, fCQL [36, 49] enables users to classify entities from a database by selecting one of pre-determined classification contexts stored in the context model of a database.

The classification language fCQL is based on SQL. The clauses are adjusted to classification purposes. Instead of the *select* clause, the name of the atttribute to be classified is written in the *classify* clause. The *from* clause is the usual SQL clause stating in which table(s) data should be sought. Instead of *where* clause this approach has *with* clause which specifies condition for a classification to a particular output class. For municipalities classification the query for the class *high winter road maintenance needs* is:

CLASSIFY municipality_name
FROM municipalities
WITH length of roads is high AND number of days with snow is high

In this section, the emphasis is put on the classification by query approach based on the GLC (2.9). The logical operator $\otimes_{i=1}^{n}$ is reduced to the *and* operator to describe the AND connective in IF part of a rule [19]:

$$where \ \oplus_{j=1}^{m_k} \otimes_{i=1}^{n} A_i \circ L_i^j, \quad k \in K \tag{4.31}$$

where n is the number of input variables, K denotes the number of output classes ($K \leq R$, where R is number of rules), $\oplus$ stands for maximum operator that merges those m_k IF parts which have common THEN part (same output class $k, k = 1, ..., K$).

Finally, the query structure is extended by the suggested clause *classify_into*. This clause specifies the name of the output class to which selected tuples are classified.

A tuple in fuzzy classification activates more than one output class with different membership degrees, if it partially satisfies more than one fuzzy query. The overall output is calculated by the aggregation of all coefficients of classes, which the tuple activates, severed by activation degrees (matching degrees to respective queries) using the following equation [19]:

$$y^0(x) = \sum_{r=1}^{R} \mu_{C^r}(x) G^r(x) \tag{4.32}$$

where R is the number of classes (output fuzzy sets), $\mu_{C^r}(x)$ is the membership degree of tuple x to the class C^r and G^r is the coefficient describing class C^r.

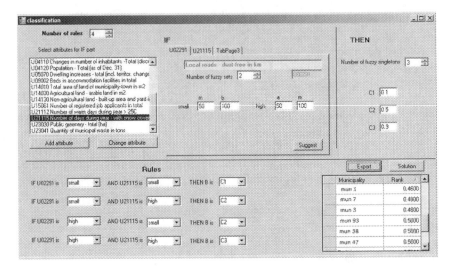

Fig. 4.19 Illustrative interface for solving classification tasks by fuzzy queries

This equation corresponds to the overall output of the Sugeno inference model expressed by weighted sum (4.30). In order to avoid this drawback, the following equation for evaluating overall output is created:

$$y^0(x) = \frac{\sum_{r=1}^{R} \mu_{C^r}(x) G^r(x)}{\sum_{r=1}^{R} \mu_{C^r}(x)} \qquad (4.33)$$

This equation corresponds to the overall output of the Sugeno inference model expressed by weighted average (4.29) and the HDM defuzzification method (4.23), applied on consequences, expressed by singletons.

Example 4.11 An institute faces a task of estimating needs for road maintenance in winter. For the sake of simplicity two input attributes are: length of roads and number of days with snow coverage again. Input attributes are fuzzified into two fuzzy sets: *small* and *high*. Output variable: needs for resources, contains three constants 0.1, 0.5 and 0.9 corresponding to singletons of *small*, *medium* and *high* or coefficients in zero-ordered Sugeno model. The illustrative interface for classifying territorial units is shown in Fig. 4.19. Parameters of fuzzy sets describing length of roads are visible in interface. Creating rule base is intuitive. At the beginning, classification space is initialized by setting number of rules, output classes, database attributes and parameters of their fuzzy sets. Consequently, each rule is constructed by selecting appropriate combination of attributes' fuzzy sets and output classes.

Three fuzzy queries are created from four rules (each query for one output class):

- for output class 0.1 (small) ($k = 1, m_1 = 1$):
  CLASSIFY_INTO Small
  SELECT *
  FROM municipalities
  WHERE roads is Small and snow is Small
- for output class 0.5 (medium) ($k = 2, m_2 = 2$):
  CLASSIFY_INTO Medium
  SELECT *
  FROM municipalities
  WHERE (roads is Small and snow is High) or (roads is High and snow is Small)
- for output class 0.9 (high) ($k = 3, m_3 = 1$):
  CLASSIFY_INTO High
  SELECT *
  FROM municipalities
  WHERE roads is High and snow is High

The minimum t-norm is used as AND connective and maximum t-conorm as OR connective.

In the next step weighted sum (4.32) is applied on each selected tuple (tuples that activated at least one rule). The result is shown in tabular form in the lower right part of interface. Concerning the option for exporting results, spreadsheet software and thematic maps extend the users' abilities to analyse tasks. □

The observation that fuzzy selection and classification can be solved by the same basis (query engine) leads to the construction of a conceptual model of their integration shown in Fig. 4.20. When the user wants to query data, the process marked with the solid line is activated. When the user wants to classify data, the process marked with the dashed line is used.

4.6 Remarks to Applications

Even though the fuzziness is closely related to phenomena in social sciences and business, the mathematics of fuzzy logic (mainly fuzzy inference systems) is prevalently applied in engineering and computer science [1, 37]. This trend allows engineering systems to be more sophisticated and powerful. We could reach the same goal in social sciences and business, if we efficiently support them by fuzzy logic. The experiments have illustrated that the fuzzy control systems and their counterpart fuzzy classifiers can be used in business and policy making.

In [8] we can find simplified inference procedure, which can be used for less complex tasks and automatized, e.g. by in-house developed software. Furthermore,

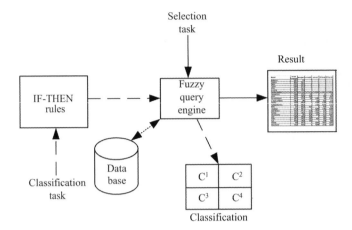

Fig. 4.20 Possible integration of fuzzy selection and classification based on [19]

CRM systems could be improved by simplified fuzzy inference or classification, which might bring benefits to both sides (company and its customers) [35].

When users are willing to accept less accurate, but flexible systems and want to include approximate reasoning, a fuzzy classification tool is the right choice and the fuzzy tool kit should be kept on the shelf [27].

Software tools focused on fuzzy inference are mainly highly parametric and complex tools in order to solve wide variety of complex tasks, but are complicated for users (domain experts which are not familiar with fuzzy logic and databases). One of possible solution is the approach based on integrating fuzzy selection and classification. Another option to avoid these disadvantages is the shell to build fuzzy inference systems in Java suggested in [30].

An interesting contribution in this direction is a newly realized survey of fuzzy systems software [2]. It can provide valuable information for businesses, public institutions and researchers that cope with the fuzziness of real world in their work and search for suitable tools.

Fuzzy inference and classification tasks can be realized frequently with different parameters. For instance, company would like to classify sellers in different regions for rewards. Rule: *if number of sold items is high and average persuasion time is small then reward is high* is generally true, but parameters of fuzzy set *high number of sold items* differs in respect to the facts such as unemployment, income, climatic conditions and the like. Hence, fuzzy sets and their parameters should be somehow stored and modified when needed. Companies databases are usually relational ones. If parameters of fuzzy sets are stored in these databases in a way that normal forms and integrity rules are preserved, then all required data are in one place. Chapter 5 is dedicated to manipulating fuzzy data (collect, store, modify, delete) in traditional relational databases.

References

1. Abdullah, M.L., Abdullah, W.S., Tap, A.O.: Fuzzy sets in the social sciences: an overview of related researches. J. Teknologi **41**, 43–54 (2004)
2. Alcalá-Fdes, J., Alonso, J.M.: A survey of fuzzy systems software: taxonomy, current research trends and prospects. IEEE Trans. Fuzzy Syst. **24**, 40–56 (2016)
3. Bardossy, L., Duckstein, L.: Fuzzy Rule-Based Modeling with Application to Geophysical, Biological and Engineering Systems. CRC, Boca Raton (1995)
4. Bezdek, J.C.: Fuzzy models and digital signal processing (for pattern recognition): is this a good marriage? Digit. Signal Process. **3**, 253–270 (1993)
5. Bilgiç, T., Türkşen, I.B.: Measurement and elicitation of membership functions. In: Pedrycz, W., Skowron, A., Kreinovich, V. (eds.) Handbook of Granular Computing, pp. 141–153. Wiley-Interscience, Chichester, West Sussex (2008)
6. Boghossian, P.: Knowledge of logic. In: Boghossian, P., Peacocke, C. (eds.) New Essays on the a priori, pp. 229–254. Clarendon Press, Oxford (2000)
7. Bouchon-Meunier, B., Mesiar, R., Marsala, Ch., Rifqi, M.: Compositional rule of inference as an analogical scheme. Fuzzy Sets Syst. **138**, 537–550 (2003)
8. Bojadziev, G., Bojadziev, M.: Fuzzy Logic for Business, Finance and Management, 2nd edn. World Scientific Publishing Co., Ltd., London (2007)
9. Branco, A., Evsukoff, A., Ebecken, N.: Generating fuzzy queries from weighted fuzzy classifier rules. In: ICDM Workshop on Computational Intelligence in Data Mining, pp. 21–28. Houston (2005)
10. Carrasco, R.A., Vila, M.A., Galindo, J.: Using dmFSQL for financial clustering. In: Chen, C.-S., Filipe, J., Seruca, I. (eds.) Enterprise Information Systems, vol. VII, pp. 113–119. Springer, Dordrecht (2006)
11. Cios, K., Pedrycz, W., Swiniarski, R.: Data Mining Methods for Knowledge Discovery. Kluwer Academic Publishers, Boston (1998)
12. Córdon, O., Herrera, F., Pelegrín, A.: Applicability of the fuzzy operators in the design of fuzzy logic controllers. Fuzzy Sets Syst. **86**, 15–41 (1997)
13. Driankov, D., Hellendoorn, H., Reinfrank, M.: An Introduction to Fuzzy Control, 2nd edn. Springer, Berlin, Heidelberg (1996)
14. Garibaldi, J.M., John, R.I.: Choosing membership functions of linguistic terms. In: 12th IEEE International Conference on Fuzzy Systems (FUZZ'03), pp. 578–583. St. Louis (2003)
15. Gorzałczany, M.: Computational Intelligence Systems and Applications. Physica-Verlag, Heidelberg (2002)
16. Enderton, H.B.: A Mathematical Introduction to Logic. Academic Press, San Diego (2001)
17. Guney, K., Sarikaya, N.: Comparison of Mamdani and Sugeno fuzzy inference system models for resonant frequency calculation of rectangular microstrip antennas. Prog. Electromagn. Res. B **12**, 81–104 (2009)
18. Gupta, M.M., Oi, J.: Theory of t-norms and fuzzy inference methods. Fuzzy Sets Syst. **40**, 431–450 (1991)
19. Hudec, M., Vujošević, M.: Integration of data selection and classification by fuzzy logic. Expert Syst. Appl. **39**, 8817–8823 (2012)
20. Hudec, M., Vujošević, M.: A fuzzy system for muncipalities classification. Central. Euro. J. Oper. Res. **18**, 171–180 (2010)
21. Hu, Y.C., Chen, R.S., Tzeng, G.H.: Finding fuzzy classification rules using data mining techniques. Pattern Recogn. Lett. **24**, 509–519 (2003)
22. Ishibuchi, H., Yamamoto, T.: Comparison of heuristic criteria for fuzzy rule selection in classification problems. Fuzzy Optim. Decis. Making **3**, 119–139 (2004)
23. Jang, J.-S.R., Sun, C.-T., Mizutani, E.: Neuro-Fuzzy and Soft Computing—A Computational Approach to Learning and Machine Intelligence. Prentice Hall, New Jersey (1997)
24. Kansal, V., Kaur, A.: Comparison of Mamdani-type and Sugeno-type FIS for water flow rate control in a rawmill. Int. J. Sci. Eng. Res. **4**(6), 2580–2584 (2013)

25. Kastner, J.K., Hong, S.J.: A review of expert systems. Eur. J. Oper. Res. **18**, 285–292 (1984)
26. Konar, A.: Computational Intelligence: Principles, Techniques and Applications. Springer, Berlin, Heidelberg (2005)
27. Kuncheva, L.: How good are fuzzy if-then classifiers? IEEE Trans. Syst. Man Cybern. Part B **30**, 501–509 (2000)
28. Lee, C.C.: Fuzzy logic in control systems: fuzzy logic controller. IEEE Trans. Syst. Man Cybern. **20**, 404–435 (1990)
29. Li, B.: Defuzzification strategy and defuzzification analysis. Ph.D. thesis, University of Aachen (1996)
30. Lopez-Ortega, O.: Java Fuzzy Kit (JFK): a shell to build fuzzy inference systems according to the generalized principle of extension. Expert Syst. Appl. **34**, 796–804 (2008)
31. Magdalena, L.: Fuzzy-rule based systems. In: Kacprzyk, J., Pedrycz, W. (eds.) Springer Handbook of Computational Intelligence, pp. 203–218. Springer, Heidelberg (2015)
32. Magdalena, L., Monasterio, F.: A fuzzy logic controller with learning through the evolution of its knowledge base. Int. J. Approx. Reason. **16**, 335–358 (1997)
33. Mamdani, E.H., Assilian, S.: An experiment in linguistic synthesis with a fuzzy logic controller. Int. J. Man-Mach. Stud. **7**, 1–13 (1975)
34. Mas, M., Monserrat, M., Torres, J.: Modus ponens and modus tollens in discrete implications. Int. J. Approx. Reason. **49**, 422–435 (2008)
35. Meier, A., Werro, N.: A fuzzy classification model for online customers. Informatica—Int. J. Comput. Inf. **31**, 175–182 (2007)
36. Meier, A., Werro, N., Albrecht, M., Sarakinos, M.: Using a fuzzy classification query language for customer relationship management. In: 31 Conference on Very large Databases (VLDB 2005), pp. 1089–1096. Trondheim (2005)
37. Meyer, A., Zimmermann, H.J.: Applications of fuzzy technology in business intelligence. Int. J. Comput. Commun. Control **VI**(3), 428–441 (2011)
38. Miller, G.A.: The magical number seven, plus or minus two. Some limits on our capacity for processing information. Psychol. Rev. **63**, 81–97 (1956)
39. Mouzouris, G.C., Mendel, J.M.: Nonsingleton fuzzy logic systems: theory and application. IEEE Trans. Fuzzy Syst. **5**, 56–71 (1997)
40. Runkler, T.A., Glesner, M.: Defuzzification with improved static and dynamic behaviour: Extended center of area. In: 1st European Congress on Fuzzy and Internet Technologies, pp. 845–851. Aachen (1993)
41. Saletić, D.Z., Velašević, D.M., Mastorakis, N.E.: Analysis of basic defuzzification techniques. In: Mastorakis, N.E., Mladenov, V. (eds.) Recent Advances in Computers, Computing and Communications, pp. 247–252. WSEAS Press (2002)
42. Siler, W., Buckley, J.J.: Fuzzy Expert Systems and Fuzzy Reasoning. Wiley, New York (2005)
43. Sugeno, M., Kang, G.T.: Structure identification of fuzzy model. Fuzzy Sets Syst. **28**, 15–33 (1988)
44. Takagi, T., Sugeno, M.: Fuzzy identifications of fuzzy systems and its applications to modelling and control. IEEE Trans. Syst. Man Cybern. **15**, 116–132 (1985)
45. Trillas, E.: On what I still hope from fuzzy logic. In: Seising, R., Trillas, E., Kacprzyk, J. (eds.) Towards the Future of Fuzzy Logic, Studies in Fuzziness and Soft Computing, vol. 325, pp. 31–54. Springer, Heidelberg (2015)
46. Trillas, E., Moraga, C.: Reasons for careful design of fuzzy sets. In: 8th Conference of the European Society for Fuzzy Logic and Technology (EUSFLAT 2013), pp. 140–145. Milan (2013)
47. van Leekwijck, W., Kerre, E.E.: Defuzzification criteria and classification. Fuzzy Sets Syst. **108**, 159–178 (1999)
48. Verkulien, J.: Assigning membership in a fuzzy set analysis. Sociol. Methods Res. **33**, 462–496 (2005)
49. Werro, N., Meier, A., Mezger, C., Schindler, G.: Concept and implementation of a fuzzy classification query language. In: International Conference on Data Mining (ICDM 2015), pp. 208–214. Las Vegas (2005)

50. Wu, D., Mendel, J.M., Joo, J.: Linguistic summarization using if-then rules. In: 2010 IEEE International Conference on Fuzzy Systems, pp. 1–8. Barcelona (2010)
51. Yager, R.R.: Knowledge-based defuzzification. Fuzzy Sets Syst. **80**, 177–185 (1996)
52. Zadeh, L.A.: From computing with numbers to computing with words—from manipulation of measurements to manipulation of perceptions. Int. J. Appl. Math. Comput. Sci. **12**, 307–324 (2002)
53. Zadeh, L.A.: Outline of a new approach to the analysis of complex systems and decision processes. IEEE Trans. Syst. Man Cybern. **SMC-3**, 28–44 (1973)
54. Zeng, K., Zhang, N.Y., Xu, W.L.: A comparative study on sufficient conditions for Takagi-Sugeno fuzzy systems as universal approximators. IEEE Trans. Fuzzy Syst. **8**, 773–780 (2000)
55. Zimmermann, H.J.: Fuzzy Set Theory—and Its Applications. Kluwer Academic Publishers, London (2001)

Chapter 5
Fuzzy Data in Relational Databases

Abstract Many agree that relational databases, like any other model of the real world, are imperfect artefacts. Hence, they cannot cover all occurrences and variety of data, in our case fuzzy data. Fuzzy values of attributes cannot be directly stored in traditional relational databases due to the first normal form. On the other hand, relational databases are broadly used. We firstly examine the way, how to store fuzzy data in traditional relational databases by satisfying normal forms in order to keep the integrity of a database in an usual way. Databases which are in use should be straightforwardly converted into fuzzy relational databases, when users decide that some attributes are better expressed by fuzzy data than by crisp values. Moreover, attributes which remain crisp, should not be affected. This improvement can be realized by fuzzy meta model of relational database. The second part of the chapter is focused on querying and summarizing fuzzy databases.

5.1 Classical Relational Databases

Relational databases are widely used in commercial world and governmental agencies for storing data. The normalization is a crucial process in database design in order to achieve a consistent database capable to ensure the integrity of the collected data. A database is considered to be normalized, if it is minimally in the third normal form [13, 30].

The normalization is a process of efficiently organizing data in a relational database. It means that redundancy is eliminated, that is, each value is stored in one place and used in relationships, if needed. Another aspect is that data dependencies make sense, i.e. table (relation) consists of related data only.

Database is in the first normal form (1NF), when the value of any attribute is a single value from the domain of the attribute. It implies that a subset of several values is not allowed.

The crucial element in a relation is the primary key, i.e. one or several attributes which unambiguously identify each record in that relation. Database is in the second normal form (2NF), when it is in the 1NF and each non-key attribute is fully functionally dependent on the primary key. It means that non-key attributes are

© Springer International Publishing Switzerland 2016
M. Hudec, *Fuzziness in Information Systems*,
DOI 10.1007/978-3-319-42518-4_5

not dependent on a part of the primary key, when primary key consists of several attributes.

Database is in the third normal form (3NF), when it is in the 2NF and each non-key attribute is not transitively dependent on the primary key.

The primary key (also mentioned in Sect. 2.1) is an unique value consisted of one or several attributes which clearly identify the whole tuple. Hence, all attributes included in the primary key have to be crisp values. In other words, the ambiguity in the primary key is out of question.

Example 5.1 In a database describing students let us have the relation STUDENT described with attributes #id, name, date of birth, age, telephone contact. All attributes are precise ones. Furthermore, the relation STUDENT is not in a 1NF, because the student may have more than one telephone number (e.g. two mobile and one landline). Herewith, these values cannot be simply written in the column telephone_contact. In order to keep the database consistent, we should create additional relation TELE-PHONE_NUM and properly link it with the STUDENT table. The relational model is shown in Fig. 5.1.   □

The difference between relation and relationship should be emphasized, because sometimes they are improperly used in database literature. The relation does not explain, how are two tables "connected" (e.g. STUDENT and TELEPHONE_NUM in Fig. 5.1). The relation stands for a table, e.g. STUDENT in the same figure. All rows represent the subset of the Cartesian product. Strictly speaking, the table is a graphical or user-friendly representation of a relation. The relationship explains the connection between tables by primary—secondary keys.

Imprecision in attributes' values can be, in a limited way, expressed without fuzzy logic. This is achieved by the NULL value introduced in [12] and further developed in [11]. The NULL value of an attribute A indicates that a value for a particular tuple could be any value from the domain D, including non-applicable one. Any comparison with NULL value in database queries creates an outcome that is neither true (1) nor false (0) called *maybe*. In terms of the Łukasiewicz three-valued logic, term *maybe* represents the truth value of 0.5 [5].

Storing fuzzy data in any relational database management system (RDBMS) seems to be impossible due to the 1NF. The 1NF states that only atomic values of attributes can be stored in relational tables, i.e. "none of its domains has elements which are themselves sets" [13]. It means that, if the database attribute is defined

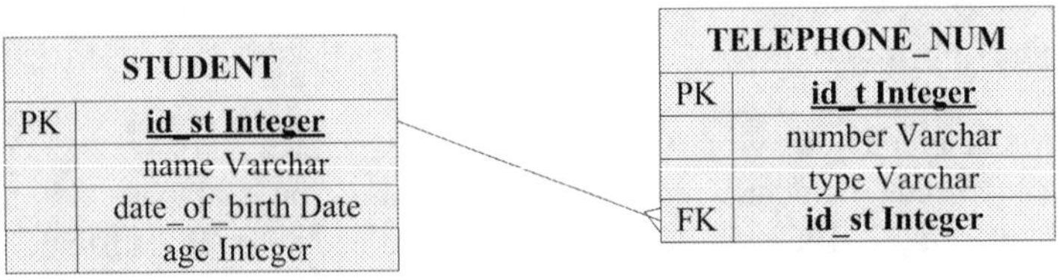

STUDENT			TELEPHONE_NUM		
			PK	**id_t Integer**	
PK	**id_st Integer**			number Varchar	
	name Varchar			type Varchar	
	date_of_birth Date		FK	**id_st Integer**	
	age Integer				

Fig. 5.1 Relationship STUDENT—TELEPHONE_NUM satisfying the 1NF

over a domain of real numbers, e.g. pollution, then we cannot express it as a triangular fuzzy number (Fig. 1.5), because we need three real numbers which in fact represent a set. Moreover, these three values are not independent, because only in proper order they represent triangular fuzzy number (1.33). In the case of relation TELEPHONE_NUM shown in Fig. 5.1, the order of telephone numbers is irrelevant.

If these parameters are stored in a way that requirements of normality are satisfied, then we can manage fuzzy data in a traditional relational database.

5.2 Fuzziness in the Data

In databases many data are precise ones, e.g. number of passengers and number of sold items. On the other hand, many other data are stored and used as numbers which pretend precision. These data are fuzzy either in their nature, or caused by the non-ideal instruments for measurements (tolerance interval). The fuzziness is amplified, when both types appear. Data are also fuzzy, when they represent estimations of people. For example

> environmental data, quality of life data and measurements of continuous one-dimensional quantities cannot be adequately expressed by crisp numbers [42].

A good example of the first type is the flooded level marked on wall illustrated in Fig. 5.2, where we cannot clearly state, which crisp value is the best option.

Examples of the second type are values measured by instruments. We should keep in mind that the measurement made by a measuring instrument is usually approximate due to the tolerance interval [27]. It means that the crisp real value is somewhere in the (small) interval $[a, b]$, but we do not know exactly where, i.e. $\mu(x) = 1$ for $x \in [a, b]$ and $\mu(x) = 0$ for $x \notin [a, b]$.

Further imprecision arises from people' estimation, e.g. observations or answers in questionnaires. For instance, someone could declare that the speed was approximately 90 km/h, but for sure not lower than 75 km/h and not higher than 110 km/h. This uncertainty can be managed by triangular or trapezoidal fuzzy number.

Even when data are expressed as linguistic terms, problem of choosing the right term may occur. Let us, for example, express the number of days with snow coverage attribute by linguistic terms according to the partition shown in Fig. 1.17. Value 25 is expressed as linguistic term *small*, but value 143 (crossover point or maximal

Fig. 5.2 Flooded level
marked on wall

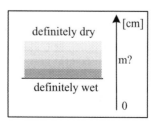

uncertainty point between sets *small* and *medium*) may be expressed as set of terms {small, medium}, because we are not sure which term is a better option.

The relational databases, like any other model of a real world, are imperfect artefacts [30]. Hence, they cannot cover all occurrences and variety of data including fuzzy data. If the real world is crisp, then classical relational databases would perfectly meet all requirements. But the real world is not deterministic. It means that the values of attributes are not known with sufficient precision to justify the use of traditional databases to store these data.

Chapter 4 is dedicated to fuzzy inference. For example, expert(s) determine(s) parameters of fuzzy sets in IF-THEN rules. These sets should be efficiently stored for further adjustment and use in inference tasks. The PERT diagrams are widely used in project managements. Each activity is expressed by three numeric values (optimistic, expected and pessimistic) instead of just one crisp value. Estimation of activity duration and time of completion is often uncertain. Fuzzy logic helps, because it covers this uncertainty [4, 44]. Fuzzy data for these tasks should be also efficiently stored in a structured way and available by request.

5.3   Fuzzy Databases: An Overview

A fuzzy database is a database capable to store fuzzy information related to tuple (row in a database), attribute (column in a database) or objects in a database [8]. Presumably, the term fuzzy database was introduced in [20]. In fuzzy databases, basic model, similarity based models, possibility based models and GEFRED model are the main approaches. In order to provide overview of fuzzy databases, these models are roughly explained.

- Basic model [30] is the simplest one. An attribute (column) expressing uncertainty of each tuple is added to the relation. The domain of this attribute is the unit interval. Though this model is the simplest and has some limitations, it can be straightforwardly added to any relational DBMS. The basic model is discussed in this chapter.
- Similarity based model or the Buckley-Petry model [6, 7] utilizes similarity relations [46] in the relational model. The value of attribute is taken from the finite set of numbers or labels. If we are not sure which value is the best for a particular tuple, we can write both, e.g. *very small* and *small*. Furthermore, similarity threshold can be added in order to get values the similarity of which is larger than the threshold value.
- Possibility models utilize the possibility theory [45] in databases. The knowledge about values of an attribute A (on a domain D) of a tuple r can be represented by the possibility distribution $\pi_A(r)$ on the domain D. The most important possibilistic models are [16]: Prade-Testemale model [31], Umano-Fukami model [40] and Zemankova-Kandel model [47]. These models differ in ways of expressing possibilities for unknown, non-applicable and undefined values.

- The GEFRED model (Generalized Fuzzy Relational Database) [26] is a synthesis of aforementioned models. It is based on generalized fuzzy domains of attributes and fuzzy relationships capable to handle a wide range of fuzzy information. This model was expanded with fuzzy division [14] and server for queries [15].

Recent overview of fuzzy databases can be found in [16, 24].

Except the basic model, these models are sophisticated enough to cope with variety of fuzziness in attributes, relationships between entities and flexible queries. However, they lack broad real applications due to their complex structures and lack of commercial tools unlike the traditional relational databases. We should not neglect the fact that users are familiar with classical relational databases and prefer to continue working in the same way, if it is only possible. Further, database designers expect clear methodology supported by available CASE tools. The following statement [36] explains the main drawback of fuzzy databases.

> Although this area has been researched for a long time, concrete implementations are rare. Methodologies for fuzzy-relational database applications development are nonexistent.

Undoubtedly, sophisticated fuzzy database models have their advantages for complex tasks, for example, in research of complex socio-natural phenomena. On the other hand, companies (especially small and medium sized enterprises) would prefer a simple to use database tool, as is the case of classical databases. That is why, aforementioned sophisticated models are not further examined.

To summarize, main reasons, why realizations of fuzzy databases lag behind the theory, presumably are the following:

1. The research and applications of managing fuzzy data in databases have been hampered by the fast development of different database technologies. Fuzzy approach should be adapted to each paradigm or at least main paradigms.
2. When we focus our interest on relational databases, nowadays broadly used, we find out that users are accustomed to clear methodology and variety of CASE tools, which is not the case in fuzzy databases.
3. Users are not interested in another database in their companies. They welcome, when managing fuzzy data is working as a part of the existing database, if possible with trouble-free extension.

This section is focused on fuzzy relational database solutions, which follow these findings.

5.4   Basic Model of Fuzzy Database

This is the simplest form of a fuzzy database which can be straightforwardly created in any RDBMS.

5.4.1 Structure of Basic Model

In a basic model of fuzzy database an additional column expressing membership degree of a tuple to the relation is created. The benefit is a simple solution which does not affect normality of a database, that is, if the initial relation is in 3NF, it will also remain in 3NF after adding this column. In addition, when a relational database is migrated to the fuzzy one, such a column has to be added only into relations which store uncertain tuples. The drawback is in unclear meaning of the membership degree. This degree might tell us the uncertainty of a tuple as a whole (e.g. interviewer's opinion, about how familiar was the respondent with the topic of questionnaire), the uncertainty of a specific attribute (e.g. respondent is fully willing to cooperate, but is able to give only roughly estimated value), or the uncertainty expressing influences among attributes.

The possible solution is adding an attribute, where uncertainty is described as open text. Nevertheless, new issues appear: (i) for user it might be a burdensome task to express uncertainty by a short and more or less uniform sentence and (ii) it is somewhat a complicated task for a query engine to read and properly understand the meaning of these sentences.

Example 5.2 Let us have relation BUILDING with attributes #id, age, size, suitability_living, energy_consumption. During data collection and storing data in database users have noticed that values for some buildings are estimated or even guessed. Thus, the existing relation BUILDING is extended into the BUILDING(#id, age, size, suitability_living, energy_consumption, m), where attribute $m \in (0, 1]$ stands for the membership degree. The interval is open from the left side, because tuples, which clearly do not belong to the relation, are not considered. This degree might be related to age, suitability for living or estimated energy consumption. For someone who just writes these values into the table, mining might be clear, but later even the same user might forget it.

In order to avoid this problem, the additional attribute: explain_m is included. Hence, two tuples might be expressed as follows: (55, 120, 700, good, 4230, 0.8, age is uncertain—no information from cadastre, but specific architectonic parts indicate the date, when house had been built) and (56, 90, 823, medium, 4235, 0.6, suitability for living is more medium than good).   □

The basic model can be efficiently used in the M:N relationship. An example is relationship between tables EXPERT and FIELD_OF_EXPERTISE (Fig. 5.3) expressed in the notation of Chen [10]. A relational database is not capable to directly manage M:N relationship. The solution is a bridge table shown in Fig. 5.4. This table

Fig. 5.3 The M:N relationship between relations EXPERT and FIELD OF EXPERTISE

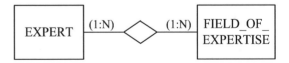

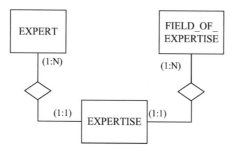

Fig. 5.4 The M:N relationship from Fig. 5.3 realized by bridge table

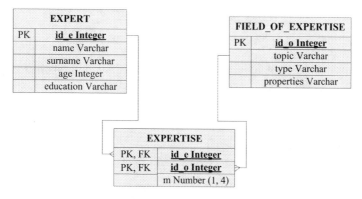

Fig. 5.5 The relationship between tables EXPERT and FIELD OF EXPERTISE and their respective attributes

is suitable for adding membership degrees. In the bridge table, degrees express suitability of an expert E_p for the field of expertise F_q. The relational model of three tables from conceptual model (Fig. 5.4) is depicted in Fig. 5.5. In this way, the meaning of the membership degree is clearer in comparison with the basic model described in the Example 5.2.

5.4.2 Querying Basic Model

In crisp or fuzzy query the additional column expressing uncertainty should be considered. Query creates a subrelation over the relation in the database, i.e. part of relation is selected. Hence, the selected tuple cannot belong to the subrelation with a higher degree than it does to the initial relation. It holds when m is not subject of the query condition. The overall query matching degree is expressed as:

$$\mu_{A_Q}(r) = t(\mu_Q(r), m) \tag{5.1}$$

where $\mu_Q(r)$ stands of a query satisfaction degree for tuple r and t is a t-norm. If query is expressed by crisp condition, instead of $\mu_Q(r)$, value of 1 is used for selected tuples.

For relations, which do not have column m, this query works as usual fuzzy query discussed in Chap. 2, because $m = 1$ for each tuple in these relations.

Example 5.3 Let us have relation BUILDING from Example 5.2 shown in Table 5.1. The following queries are realized:
(i) select buildings belonging to the relation with degree higher or equal 0.75;
(ii) select buildings where age is high;

The query (i) is an usual SQL query of structure:
SELECT *
FROM building
WHERE $m \geq 0.75$
The * means that all columns are selected. The selected subrelation contains all buildings except buildings $b5$ and $b6$.

The query (ii) is a fuzzy SQL query of structure:
SELECT id, age, m
FROM building
WHERE age is high

If the term *high* is expressed as R fuzzy set (1.23) with parameters $a = 90$ and $m = 95$, then the selected buildings ranked downward from the best to the worst tuple are in Table 5.2. The minimum function was used as t-norm between the query satisfaction degree and column m.

If column m is a subject of query condition, then tuple might belong to query with a higher degree than it did to the initial relation in a database. An example is the query which selects buildings that slightly belong to the relation BUILDING. If term slightly is expressed as L fuzzy set (1.22) with parameters 0.15 and 0.4, then $b6$ meets query with degree of 1 and $b5$ with degree of 0.2.   □

In the same way queries on the bridge table are realized. In this case the *join* clause has to be included.

Table 5.1 Relation BUILDING in the basic model of fuzzy database

#id	Age	Size (m^2)	suit_living	en_consumption	m
b1	98	380	Good	4290	0.90
b2	89	390	Excellent	3250	1.00
b3	91	400	Low	8320	0.75
b4	125	280	Medium	4280	0.85
b5	68	410	Good	5290	0.35
b6	70	382	Excellent	51030	0.12

Table 5.2 The result of the query condition *age is high* realized on relation BUILDING

#id	Age	m	$\mu_Q(r)$	$\mu_{A_Q}(r)$ (5.1)
b1	98	0.90	1	0.90
b4	125	0.85	1	0.85
b3	91	0.75	0.20	0.20

Example 5.4 The task is to find older experts for a demanding task in field of expertise *B* with high score of suitability. Thus, the query is of structure:

SELECT expert.name, expert.surname, field_of_expertise.topic
FROM field_of_expertise INNER JOIN (expert INNER JOIN expertise ON
expert.id_e=expertise.id_e) ON field_of_expertise.id_o=expertise.id_o
WHERE field_of_expertise.topic='B' AND expertise.m > 0.8 AND expert.age is
high

At the beginning the parameters of fuzzy set *high age* should be defined. In the second step, all tuples which fully or partially meet the condition are selected (e.g. by procedure examined in Sect. 2.3). Finally, the query matching degree is calculated by (5.1). □

Suitable t-norm for aggregating $\mu_Q(r)$ and *m* in (5.1) is minimum t-norm (1.47). Let us have tuple which meets query condition with degree of 0.5 and belong to the relation with degree of 0.5. We expect that the answer is 0.5, which is calculated only be minimum t-norm. Concerning the *and* connective among atomic predicates in compound query condition, $\mu_Q(r)$ can be calculated by different t-norms following the nature of query condition. This topic is discussed in Sect. 2.4.

5.5 Fuzzy Data in Traditional Relational Databases Managed by Fuzzy Meta Model

Storing fuzzy data in the classical relational database seems to be impossible at the first glance, due to the 1NF. To recapitulate, the 1NF states that only atomic values of attributes can be stored in relational tables, that is, none of attributes has values which are themselves sets [13].

The main-stream in developing fuzzy databases suggested that the requirement of the 1NF must be weakened, e.g. [30]. According to Ma and Yan [24], the normalization theory of the classical relational database model must be relaxed in order to solve the problems of anomalies and redundancies that may exist in non-normalized fuzzy relational databases and to provide a theoretical guideline for database design.

Fuzzy functional dependencies are generalization of classical functional dependencies [33, 37, 43]. Based on the fuzzy functional dependencies, normal forms such as Fuzzy First Normal Form, q-Fuzzy Second Normal Form, q-Fuzzy Third Normal

Form are constructed [9]. As a result, q-keys, Fuzzy First Normal Form, q-Fuzzy Second Normal Form, q-Fuzzy Third Normal Form, and q-Fuzzy Boyce-Codd Normal Form are formulated. Furthermore, lossless-join decompositions into q-F3NFs are discussed in [25].

On the other hand, keeping the classical 1NF, 2NF and 3NF satisfied brings also benefits. The reasons are: (i) any traditional relational database management system can be used without violation of classical integrity rules; (ii) migrating existing database into fuzzy one may keep already met requirements for normality; (iii) methodology for design of traditional relational databases is solid and supported by CASE tools.

There are many real situations, where relationships between relations are crisp, but only values of several attributes are fuzzy. In this case, the adaptation of traditional relational databases is a rational option.

In order to manage fuzzy data in classical relational databases, Škrbić [34] suggested the relational fuzzy meta model. In this way relational model for managing entities and relational fuzzy meta model for managing fuzzy data of these entities are merged. If parameters of fuzzy sets are stored in a way that requirements of normality and integrity are satisfied, then we can manage fuzzy data in any RDBMS. In the basic model of fuzzy database (Sect. 5.4) the support for imperfect information is at the tuple level. In this approach the support of imperfect information is on the attribute level, clearly stating which attribute and in what way is affected by fuzziness for each tuple.

5.5.1 Creating Fuzzy Meta Model

Each fuzzy data or linguistic term is represented by membership function (usually several parameters, except for the singleton fuzzy set) (Sect. 1.2.2). Therefore, the relation (table) TRIANGULAR contains attributes #fuzzy_id, a, m, b, in order to manage manipulation of terms expressing *medium* value or *similar to m*. In the same way other relational tables for storing other types of fuzzy sets are created: TRAPEZOIDAL(#fuzzy_id, a, m1, m2, b), L_FUZZY(#fuzzy_id, m, b) and the like. Singleton fuzzy set contains only one parameter. It may tempt us to store this value in a relational model of entities as a precise value, instead of storing it in a fuzzy meta model. But it is a wrong decision, due to reasons explained later on.

The opinions can be expressed by terms from the set of ordered linguistic terms. This option is used in questionnaires, among others. The frequently used scale for measuring opinions is Likert's [21] consisted of a set of odd number of linguistic terms (equal number of positive and negative terms and one neutral) related to the question asked. This scale can be managed by fuzzy logic [22]. The question in questionnaire represents fuzzy variable, whereas possible answers (terms) are linguistic labels defined by syntactic and semantic rules (Sect. 1.4). Users cannot always unambiguously choose one term form the set of the ordered terms. Answer may be like:

more negative than neutral. In terms of fuzzy set it is expressed as, e.g. {(negative, 0.75), (neutral, 0.25)}.

We would like to emphasize that Gaussian function should be avoided, because membership degrees appear on the whole domain (asymptotically approaching the value 0). Hence, it seems that imprecise data cover the whole domain and inappropriately affect functions in SQL queries, such as SUM (total) and AVG (average value).

The fuzzy meta model for extending relational database is introduced in [34]. That model is adjusted to fuzzy values used in this book and shown in Fig. 5.6. Two main tables are: IS_FUZZY stating, which attributes from all relational tables are considered as fuzzy, and FUZZY_LINK linking the fuzzy attributes from relational model of real entities to fuzzy meta model, where values of these attributes are stored. Table FUZZY_TYPE allows us to express type of fuzzy set for each tuple (triangular, trapezoidal, linguistic term, etc.).

Keeping in mind these observations we could create fuzzy meta model for a particular database. The idea of fuzzy meta model is demonstrated on the database which merges municipal statistics and data related to buildings.

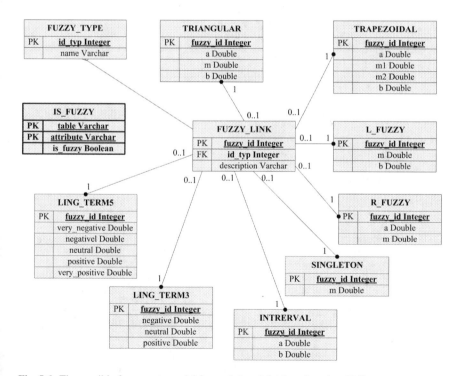

Fig. 5.6 The possible fuzzy meta model for a relational database based on [34]

In the Fig. 5.7 the fuzzy meta model (Fig. 5.6) attached to the relational model is shown. In the table MUNICIPALITY attributes pollution and opinion of inhabitants about question Q (opinion_q) do not store real values, but foreign keys to the respective tables, where these real values are stored as fuzzy. The same holds for attributes age and flooded in the table BUILDING. In this way the database is capable to store fuzzy data and meet integrity rules mentioned above.

Difference between singleton fuzzy set and crisp attribute should not be neglected. A non-fuzzy attribute is always crisp, whereas fuzzy attribute could contain crisp values for some tuples. Because in a relational table the fuzzy attribute contains only a foreign key, the real crisp value should not be stored in entities relations, but in the relation managing singletons.

Regarding the size of such overblown database, both the size of structure and the size of stored data are affected. Concerning the size of structure, it fully depends on the chosen number of fuzzy set types supported by the fuzzy meta model. In model shown in Fig. 5.7, 10 tables containing 32 attributes are added. The size of stored data in a database is affected by the number of fuzzy attributes (columns) and types

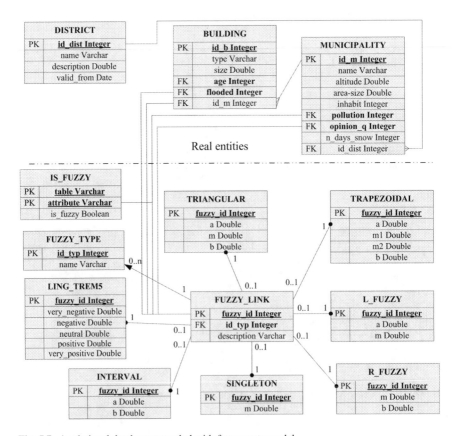

Fig. 5.7 A relational database extended with fuzzy meta model

of fuzzy sets applied on them. In the extreme situation all n attributes (excluding primary keys) for all m entities are fuzzy of the trapezoidal form. Therefore, the size of database is $9(m \cdot n)$. Anyway, user has to carefully decide which attributes should be stored as fuzzy and in which form. Keeping in mind this deduction, we are able to say that the size of fuzzy relational database could be significantly smaller than this extreme situation.

It is worth noting that a relational database extended with the fuzzy meta model is a fully relational one without weakening of any normal form and integrity rules. Theoretically, any CASE tool capable to generate Data Definition SQL code from the created model can be used. The DDL (Data Definition Language) is part of SQL focused on the creation, modification and removing the structures in which the data are going to be stored. Furthermore, it includes statements for defining indexes and other relevant procedures for working with data structures. From the practical point of view, the CASE tool should also support definition of relevant parameters for membership functions. Presumably, the first realization of CASE tool for fuzzy databases is the tool developed in Java [36].

5.5.2 Storing and Representing Tuples

Storing new tuples in the database is expressed by the following algorithm [19]:

---

check whether attribute appears in table IS_FUZZY
if attribute appears (is fuzzy = true) then
    generate ID for foreign key
    check acceptable fuzzy sets for the attribute and offer them in a combo box
    according to selected type of fuzzy set provide input boxes for parameters
    verify input values and store them in relevant tables*
else
    allow storing crisp value for attribute
end if

---

* In case of triangular fuzzy set, parameters should be verified by the rule: $D_{min} \leq a < m < b \leq D_{max}$. If $a = m = b$, then either this value should be stored as a singleton fuzzy set in the table SINGLETON (less required space and faster computations), or user should be informed to check, whether mistake appeared.

Example 5.5 A company is interested to improve customer relationship management. The first step is to collect informations from customers regarding their age, frequency of using products and opinions, how efficiently company solves complaints, by a questionnaire. Age is expected to be crisp value. Concerning the second attribute, value is expected to be estimated like: about 75 days, but not less than 70 and not more than 85. As for the third attribute, the answer could be one term from the set {very negative, negative, neutral, positive, very positive} (a Likert scale with one neutral term in the middle and two on each side). However, customers cannot always explicitly state, which linguistic term is the most suitable answer. Hence, they tend to express the answer of the type, *more negative than neutral*. A more convenient way is to allow respondents to weight linguistic terms. In this case a respondent might say: (negative, 0.75) and (neutral, 0.25) to clarify the term *more negative than neutral*.

Evidently, for storing such a data the relational database should be adjusted. The first step is creating new fuzzy meta model or adjusting existing one (if company already has it). To solve this task, we need to create tables for customers' opinions. The relational tables TRIANGULAR and LING_TERM5 (Fig. 5.6) meet the requirement for attributes of frequency of using products and of opinion about managing complaints, respectively. The attribute age is a crisp numeric one, so values are stored in the relational table CUSTOMER. Based on these assumptions, we can create table CUSTOMER consisting of attributes: #id_c (primary key), age, complaint (foreign key) and frequency (foreign key) and link them with the fuzzy meta model. Now, the database is ready for storing customers' data.   □

The next example demonstrates an interface for storing fuzzy data into the relational database of buildings and territorial units.

Example 5.6 In order to manage tuple by tuple storing data, the interface shown in Fig. 5.8 has been created. The interface is similar to interfaces of storing data in relational database. The only difference is in managing fuzzy data. User should

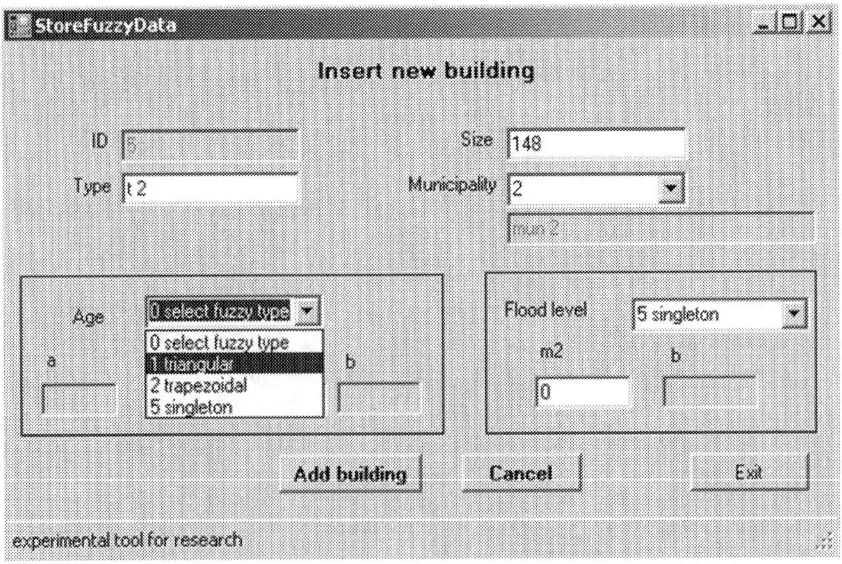

Fig. 5.8 Interface for storing fuzzy data into fuzzy relational database

select type of fuzzy set from the list of allowed fuzzy sets for each attribute and insert required parameters. The code behind interface manages validating inserted data, storing data into the right tables and link them to attributes. The procedure follows the aforementioned algorithm. □

The procedure for storing data can be automatized, if the observed data are available in a structured way. A suitable option could be an XML file, if attribute values which are fuzzy, are marked by special tag and properly explained. An option for

BUILDING

id_b	type	size	age	flooded	id_m
1	t 1	150	1101	2101	1
2	t 2	285	1102	2102	1
3	t 1	148	1103	2103	2
4	t 1	286	1104	2104	2

DISTRICT

id_dis	name	description	valid_from
1	dist 1	some text	1.6.2000
2	dist 2	some text	1.6.2000

MUNICIPALITY

id_m	name	altitude	area_size	inhabit	pollution	n_days_snow	opinion_q	id_dist
1	mun 1	125	13200	5385	3101	25	4101	1
2	mun 2	588	12905	4297	3102	83	4102	2
3	mun 3	589	25103	3920	3103	85	4103	2

IS_FUZZY

table	attribute	is_fuzzy
building	type	0
building	size	0
building	age	1
building	flooded	1
municipality	name	0
municipality	altitude	0
municipality	area_size	0
municipality	inhabit	0
municipality	pollution	1
municipality	n_days_snow	0
municipality	opinion_q	1
district	name	0
district	description	0
district	valid_from	0

FUZZY_LINK

fuzzy_id	id_type
1101	1
1102	5
1103	1
1104	5
2101	1
2102	5
2103	5
2104	2
3101	1
3102	1
3103	2
4101	6
4102	6
4103	6

FUZZY_TYPE

fuzzy_id	id_type
1	triangular
2	trapezoidal
3	L type
4	R type
5	singleton
6	linguistic term

TRIANGULAR

fuzzy_id	a	m	b
1101	140	150	160
1103	150	200	215
2101	92	105	110
3101	34	40	45
3102	90	100	105

TRAPEZOIDAL

fuzzy_id	a	m1	m2	b
2104	60	70	75	85
3103	85	95	100	110

LING_TERM5

fuzzy_id	very_negative	negative	neutral	positive	very_positive
4101	0	0	0.3	0.7	0
4102	0.85	0.15	0	0	0
4103	0	0.15	0.7	0.15	0

SINGLETON

fuzzy_id	m
1102	42
1104	58
2102	0
2103	0

Fig. 5.9 Fuzzy data in a relational database

extending the SDMX (Statistical Data and Metadata eXchange) format based on the XML to cover fuzzy data is introduced in [18].

Stored data can be in the structure depicted in Fig. 5.9. All fuzzy as well as crisp data can be easily reconstructed and further used. We see for example, that *Building 1* belongs to the *mun 1*, size is 150 m², age is about 150 years, but for sure not lower than 140 and not higher than 160 (e.g. we do not have precise date from the cadastre documentation, but only guess values according to visible architectonic elements), building was flooded in a way that under 92 cm wall was definitely wet and above 110 cm was for sure dry. In the same way, we can retrieve data about the territory, where this building is situated: pollution in is about 40 mg of measured pollutant, but for sure not lower than 35 and not higher than 45 and opinion about this territory is rather positive.

Apparently, an interface will be more suitable than looking at tables. The interface capable to meet this goal is shown in Fig. 5.10. Displaying crisp values is not an issue. These values are simply written. User sees that the altitude of chosen municipality is 588 m. Concerning fuzzy data, it is a bit complex task. Three text boxes for displaying parameters a, m and b for triangular fuzzy set are not satisfactorily explainable. Hence, a short sentence to support fuzzy data explanation should be created. An example of such a sentence is shown in Fig. 5.10 for the attribute pollution. Another option is displaying fuzzy data in graphical way like in the figures in Sect. 1.2.2.

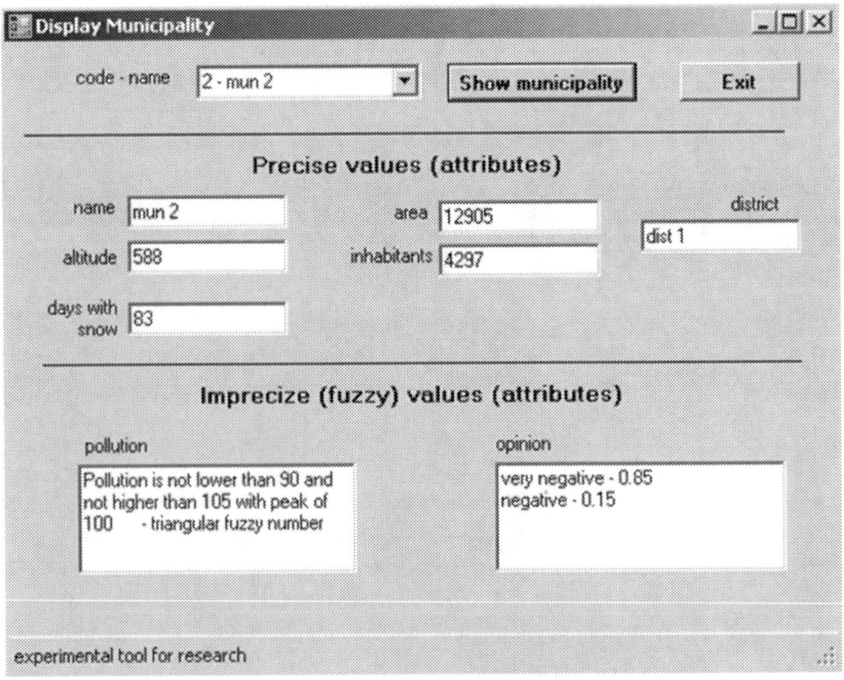

Fig. 5.10 Interface for displaying stored fuzzy data

Concerning the attribute opinion, all terms with membership degree greater than 0 are shown together with their respective degrees.

5.5.3 Inserting Fuzziness into Existing Databases

From the database designers' and users' point of view, the seamless migration into the fuzzy database is a valuable feature. In case of managing fuzziness by the fuzzy meta model, adding fuzzy data into an existing relational database can be straightforwardly realized, when required. At the beginning, user should decide which attributes will be further collected as fuzzy. For example, after some period of using a relational database, many complaints about stored values for pollution and flooded level were recorded. In order to solve this problem, the decision is migration from the classical to the fuzzy relational database.

Previously collected data of fuzzy attribute need to be migrated to the singletons (fuzzified by 4.26). Apparently, this holds only for the attributes which are going to be managed as fuzzy in the improved database. Because in a relational table the fuzzy attribute contains only foreign key, the previously collected real value is shifted to SINGLETON table in the fuzzy meta model. An unique value is generated to replace data in the relational table with the value of foreign key. These keys point to the table SINGLETON, where data are migrated. At this point the database is migrated to the fuzzy one and is ready to store fuzzy data, when new collection is launched.

The algorithm for the migration from existing relational database into fuzzy one is not complicated, but the procedure for migrating might take a lot of time, depending on the database size. In order to use this algorithm, the fuzzy meta model should be designed and all tables ready to use.

Shortly, the algorithm for migration is of the following structure:

```
select all rows from IS_FUZZY table into data reader DR
while DR is not EOF
  read name of attribute and table (T) where this attribute is
  open table T
  while T is not EOF
    read value of chosen attribute and store in temporary variable V
    generate value for new foreign key
    write this key value instead of the attribute's real value
    add row in table FUZZY_LINK and store this key value and choose singleton
    add row in table SINGLETON and fill with generated key value and
      attribute's value from V
  end while
end while
```

It is worth noting that this migration can be also realized when fuzzy database is already in use. In this case only values of chosen attribute(s) are migrated to the fuzzy meta model. Hence, consulting other rows in table IS_FUZZY is not necessary.

Example 5.7 In the database expressed by Fig. 5.9, we see attribute number of days with snow coverage (n_days_snow). Crisp values used to be collected for this attribute. But we know that precise measuring of this value is disputable, due to vagueness in this attribute: does the coverage mean a layer of snow on the whole surface of municipality during the whole day, or at least during half a day? Someone can declare that an estimated crisp value is sufficient. Nevertheless, in some tasks it might cause problems. Municipalities *mun2* and *mun3* have very similar values. Let us consider rule: *if coverage is $\geq$ 85 then government provides additional support for the winter road maintenance to the municipal council.* In this case it is obvious that the solution does not reflect reasoning that similar municipalities should receive similar support. A possible solution for this issue is fuzzy classification, where belonging to consequence or output class is a matter of degree, i.e. similar municipalities have similar membership degrees to overlapped classes (Sect. 4.5). Anyway, we cannot always rely on fuzzy classification. Hence, keeping values of this attribute as fuzzy in fuzzy relational database is a rational option.   □

When more detailed information about vague observations is required, the α-cuts (1.15) are suitable [42]. They are able to express both convex and non-convex fuzzy sets. Let us consider number of days with snow coverage again. From Fig. 5.11 we conclude that snow coverage between 52 and 127 days appeared, although in some days the coverage was very low. Between the days 56 and 98 coverage was visible, between the days 68 and 92 coverage was significant, between the days 73 and 89 high and between the days 76 and 79 very high. This information is not possible to be covered by trapezoidal fuzzy set.

Fig. 5.11 Number of days with snow coverage expressed by α-cuts and non-convex fuzzy set

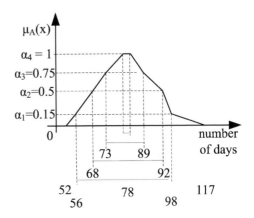

5.5.4 Managing Fuzziness in Data and in Inference Rules by the Same Database

Fuzzy rule-based systems consist of knowledge base and fuzzy inference. These topics were examined in Chap. 4, where focus on knowledge base was on the conceptual level. Topic which was not covered until now is an efficient way for storing and updating parameters of linguistic terms appearing in antecedent and consequent parts of rule base.

Attributes' values of tuples (territorial units, customers, etc.) are usually stored in a structural way, for instance in relational databases. These values are exported from a database and converted into proper format for fuzzy inference systems. Attributes' values are crisp or fuzzy. In case of former, these values are fuzzified by e.g. (4.26), if required. In case of latter, these values are directly imported into fuzzy inference systems. Fuzzy values of attributes can be stored in fuzzy meta model of a relational database as was demonstrated above.

When rule base is of the structure:

- IF pollution is high and flooded level is high, THEN reimbursement is high
- IF pollution is low and flooded level is low, THEN reimbursement is low

and fuzzy values of input values are stored in fuzzy meta model shown in Figs. 5.6 and 5.9, then the following question appears: can we store parameters of linguistic terms for rules in the same database? The answer is, yes. Moreover, this way is a reasonable option, because we benefit from two aspects [19]. Firstly, we can use the same procedures for storing and updating fuzzy data of observations and fuzzy data of linguistic terms in rules. Secondly, managing fuzzy data of entities and fuzzy sets of fuzzy rules in the same database could lead to integration of information system and fuzzy inference system.

The database model focused on managing parameters of linguistic terms in a fuzzy rule base is plotted in Fig. 5.12. Table FUZZY_TYPE contains all acceptable types of fuzzy sets. Contrary to fuzzy data of observations, linguistic terms in a rule base can be also expressed as Gaussian fuzzy set. Similarly, other types can be added, if needed. The difference in this part is the 1:N relationship, because linguistic variable contains several terms (e.g. Fig. 1.17).

Example 5.8 Let us have a rule base for managing reimbursement for flooded houses, considering their sizes. Thus, input variables are size and flooded level. Both variables are fuzzified into three fuzzy sets: *small*, *medium* and *high*. Reimbursement is fuzzified into five fuzzy sets: *very small*, *small*, *medium*, *high* and *very high*. The rule base is shown as a decision table in Table 5.3. The instance of fuzzy meta model for managing rules is shown in Fig. 5.13.

We have noticed that additional column (id_att) in tables for expressing particular fuzzy sets is required. Hence, relations in fuzzy meta model shown in Fig. 5.6 should be extended with this column. When a row stores values for inference rules, then this column is filled with foreign key of respective attribute. Otherwise, this column remains empty. Attribute fuzzy_id is a key which is generated by procedure for

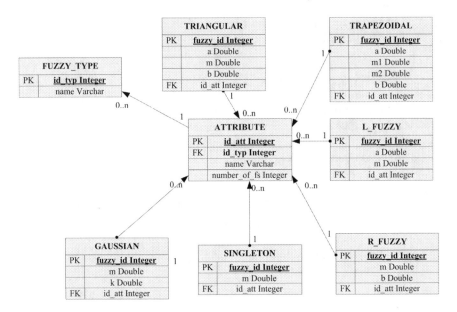

Fig. 5.12 Part of database for managing fuzzy rules

Table 5.3 Decision table for managing reimbursement

Size/Flooded level	Small	Medium	High
Small	Very small	Small	Medium
Medium	Small	Medium	High
High	Medium	High	Very high

data insertion. For instance, designers may decide that, if this value starts with 99, the row then represents linguistic term used in fuzzy inference. Otherwise, the row stores parameters of collected fuzzy data. In table ATTRIBUTE we can see that input variable *size* is fuzzified into three fuzzy sets. Parameters of these fuzzy sets are stored in tables TRAPEZOIDAL, L_FUZZY and R_FUZZY. For illustrative purpose, family of fuzzy sets for this attribute is shown in Fig. 5.14. Output variable *reimbursement* is fuzzified into five fuzzy set of trapezoidal type in order to keep limited support for all linguistic terms.   □

Linguistic quantifiers are mainly used in quantified queries and linguistic summaries examined in Chaps. 2 and 3, respectively. These objects can be straightforwardly added into fuzzy meta model in the same way as it was demonstrated for linguistic terms in fuzzy rules.

A more complex solution for managing fuzziness is provided by Fuzzy Meta-knowledge Base (FMB) explained in [16]. This base organizes all information related to vagueness in a fuzzy database. In FMB the following three elements are stored: attributes with fuzzy processing, information about these attributes as well as other

ATTRIBUTE

id_a	name	type	number_fs
att_1	size	2	3
att_2	flooded	2	3
att_3	reimbursement	2	5

TRAPEZOIDAL

fuzzy_id	a	m1	m2	b	id_a
99002	100	140	240	280	att_1
99005	30	40	70	80	att_2
99007	0	0	100	150	att_3
99008	100	150	250	300	att_3
99009	250	300	400	450	att_3
99010	400	450	550	600	att_3
99011	550	600	700	700	att_3

L_FUZZY

fuzzy_id	m	b	id_a
99001	100	140	att_1
99004	30	40	att_2

R_FUZZY

fuzzy_id	a	m	id_a
99003	240	280	att_1
99006	70	80	att_2

Fig. 5.13 An instance of part of a fuzzy meta model, where variables for fuzzy rules are managed

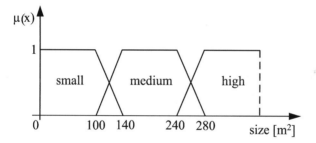

Fig. 5.14 Linguistic terms of attribute *size* in a rule base

relevant information, such as fuzzy quantifiers. Attributes with fuzzy processing are divided into 8 types. Type 1 covers crisp attributes, type 2 manages fuzzy information on domains. Both these types are also covered in fuzzy meta model. Type 3 and type 4 manage fuzziness in attributes by possibility distributions. Other four types represent variations of fuzzy degrees associated to attribute or tuple.

The FMM is suitable, when first two types of fuzziness are sufficient for explaining vagueness in the attributes, in linguistic terms of fuzzy rules and in linguistic quantifiers, as well as in cases, when vagueness should be stored in a way that classical normal forms of RDBMS are met. Furthermore, degrees associated to tuple can be added without violating classical normal forms.

5.6 Querying Fuzzy Relational Databases

In relationally structured databases (crisp or fuzzy) we can apply SQL-like queries [30]. Two well-known query languages capable to deal with fuzzy data in fuzzy databases are PFSQL [35, 38] and FSQL [16, 41].

Concerning fuzzy meta model, the basic functionality of SQL is examined in order to keep approach less demanding for tools and independent of possible variations in different relational DBMS. All further calculation is executed in the application layer adjusted to the particular database and user's needs. Algorithms and calculations discussed below are not complex, even though sometimes a bit long.

5.6.1 Aggregation Functions in Queries

Nowadays, fuzzy sets are used in many application areas such as planning, forecasting and control, among others. Functions such as, total or sum and average are widely used in this field [4]. These functions related to relational databases are discussed in [2]. This section is focused on aggregating fuzzy data by functions examined in Sect. 1.2.4.

For example, the AVG function in SQL returns the average value of a numeric attribute in the whole relation or in subrelation defined through the *where* clause of SQL. In the next two examples this function is demonstrated on fuzzy data.

Example 5.9 A historian wishes to know the average age of buildings in municipality *mun*1. SQL query has the following structure:

SELECT AVG(building.age) as average
FROM municipality INNER JOIN building ON municipality.id_m = building.id_m
WHERE id_m = "1"

When this query is applied on crisp attribute such as *size*, then the initial *avg* function of SQL is used. If the attribute is fuzzy, this function is overridden with fuzzy arithmetic function *average* created from the *sum* (1.35) and *division* (1.38) functions.

Age of building 1 is triangular number: b1.age(140, 150, 160), whereas age of building 2 is singleton: b2.age(42, 42, 42). Hence, the average is calculated as $avg(building.age) = \frac{b1.age + b2.age}{2} = \frac{1}{2}(182, 192, 202) = (91, 96, 101)$. For the second building we know age for sure, so the resulting fuzzy number has shorter support (1.9) than the support of the building 1.   □

Example 5.10 An environmentalist would like to know the average pollution in districts. Because the pollution attribute is collected on municipalities level, the query has the following structure:

SELECT AVG(municipality.pollution) AS avg_pollution, district.name
FROM district INNER JOIN municipality ON district.id_dist = municipality.id_dist
GROUP BY district.id_dist

Table 5.4 Result of query searching for average pollution in districts

District	Average pollution (mg) (A_{avg})
dist 1	35; 40; 45—triangular number
dist 2	87.5; 97.5; 100; 107.5—trapezoidal number

Table 5.5 Result from Table 5.4 shown as text description instead of fuzzy sets parameters

District	Average pollution (mg) (A_{avg})
dist 1	Is not lower than 35 and not higher than 45 with peak in 40
dist 2	Is not lower than 87.5 and not higher than 107.5 with a flat segment in interval [97.5; 100]

The averages are obtained in the same way as in previous example. Because the intent of the query is to find averages for each district, result is shown in Table 5.4. Concerning the legibility of result, short texts can replace parameters in the second column. This option is demonstrated in Table 5.5 and can be further displayed in interface in the same way as is shown in Fig. 5.10.   □

The AVG function is obtained as function SUM divided by number of considered tuples. Therefore, function SUM is not further illustrated in examples. In the same manner we can override other arithmetic functions in SQL.

Defuzzification of fuzzy averages
In Examples 5.9 and 5.10 results are fuzzy sets. These values often need to be expressed as crisp ones which represent the corresponding fuzzy values in the best way. It means that fuzzy averages need to be defuzzified. Defuzzification procedures examined in Sect. 4.2.2 can be applied. But for the case of simplicity, it can be reasonable to select value of maximal height [4]. Therefore, defuzzified value of triangular fuzzy number is

$$x_{max} = m \tag{5.2}$$

where m is the modal value.

This calculation corresponds with the defuzzified value of unimodal fuzzy set. But this defuzzification cannot be defined uniquely. Options for defuzzification of fuzzy averages are suggested in [4] as:

$$(i)\, x_{max}^{(1)} = \frac{a + m + b}{3} \tag{5.3}$$

$$(ii)\, x_{max}^{(2)} = \frac{a + 2m + b}{4} \tag{5.4}$$

$$(iii)\, x_{max}^{(4)} = \frac{a + 4m + b}{6} \tag{5.5}$$

where contrary to (5.2) the parameter 1, 2 and 4 ensures different weight for m, but also takes into consideration the values of a and b. If a triangular number is symmetric, i.e. $|b - m| = |a - m|$, then the three equations produce the same result. According to [4], usually in applications, fuzzy averages expressed by triangular numbers are in central form (symmetric). But this is not always the case. Hence, these equations can be parametrized in the following way:

$$x_{max}^{(s)} = \frac{a + s \cdot m + b}{s + 2} \tag{5.6}$$

where $s \in \mathbb{N}$. For non-symmetric triangular fuzzy numbers defuzzified average approaches the value of m, when s increases. In case of large non-symmetricity, gravity strategies, such as COG (4.21) or HOF (4.22), though a bit complex for the calculation, are better options.

Average of the trapezoidal number is calculated as the extension of (5.2):

$$x_{max} = \frac{m_1 + m_2}{2} \tag{5.7}$$

where m_1 and m_2 are boundaries of the flat segment. This equation corresponds with the COM (4.20) defuzzification method.

Correspondingly, for the non-symmetric trapezoidal average defuzzification is expressed as

$$x_{max}^{(s)} = \frac{a + s \frac{m_1 + m_2}{2} + b}{s + 2} \tag{5.8}$$

where s has the same meaning as in (5.6).

Example 5.11 Averages from the Examples 5.9 and 5.10 should be defuzzified into crisp values.

In Example 5.9 (average of building age) the result was $avg(building.age) = (91, 96, 101)$. It is obvious that this average is in the central form, hence $x_{max} = 96$.

In Example 5.10 calculated averages of pollution are in Table 5.4. Defuzzification of triangular and trapezoidal averages are calculated by (5.3) and by (5.8), where s = 3, correspondingly and shown in Table 5.6.   □

Table 5.6 Defuzzified averages of pollution for districts

District	Average pollution (mg) (A_{avg})	Defuzzified pollution (mg)
dist 1	35; 40; 45—triangular number	40
dist 2	87.5; 97.5; 100; 107.5—trapezoidal number	98.25

5.6.2 Query Conditions

If we wish to select, modify, or delete particular tuples, then we should create condition. The following four types of queries are applicable:

- crisp *where* clause on crisp attributes (classical SQL)—these query conditions are well documented in textbooks related to SQL queries
- fuzzy queries on crisp attributes—these query conditions are deeply examined in Chap. 2
- crisp queries on fuzzy attributes
- fuzzy queries on fuzzy attributes

This section is focused on the last two types of query conditions.

5.6.2.1 Crisp Queries on Fuzzy Attributes

In this kind of queries fuzzy data are compared to crisp values by traditional comparison operators, such as >, <, =. Let us firstly consider comparator equal to (=). When we fuzzify crisp value to singleton fuzzy set by (4.26), the task can be solved by possibility measure (1.28), i.e. to find possibility that fuzzy data belong to fuzzy concept. This type of query condition is illustrated in Fig. 5.15 for condition: WHERE attribute A = q, ($q \in \mathbb{R}$), for three tuples (r_1, r_2 and r_3). These tuples have fuzzy values A_{r1}, A_{r2} and A_{r3} for the attribute A consequently. We can see that tuple r_1 meets the query condition with degree $\mu_{A(r1)}$ (supremum of intersection between fuzzy data and crisp concept expressed as singleton), tuple r_2 fully meets the query condition and tuple r_3 does not meet the query condition.

Fig. 5.15 Query condition WHERE A = q, in which A is a fuzzy attribute

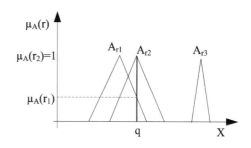

Example 5.12 The task is to select municipalities with pollution equal to 95. The condition is:

WHERE pollution = 95

Collected values of pollution are stored in a database shown in Fig. 5.9. Value 95 is fuzzified into singleton fuzzy set F as:

$$\mu_F(x) = \begin{cases} 1 & \text{for } x = 95 \\ 0 & \text{otherwise} \end{cases}$$

This singleton is in the flat segment of pollution in tuple *mun 3*. It implies that this municipality fully meets the query condition. Concerning *mun 2*, supremum of intersection is in value 0.5. The answer to query is in Table 5.7. Though the condition seems to be usual crisp where clause, fuzziness in attribute pollution is reflected in the set of selected tuples. When the crisp value of condition is in the support of fuzzy data, but not in the core, it is reflected in matching degree lower than 1.   □

Query matching degree is greater than zero, when crisp value of condition belongs to support of fuzzy data. By this simple observation we can select all tuples for calculating matching degree to query condition.

For the comparison operators $>$, $\gg$, $<$ and $\ll$ the above examined possibility measure is not the right choice, because different results are expected for conditions: *attribute A > q* and *attribute A < q*, when q is not in the core of fuzzy value. This consideration is illustrated in Fig. 5.17, where we expect that matching degree for *attribute A < q* is lower than for *attribute A > q*.

The condition: WHERE attribute A $< q$ ($q \in \mathbb{R}$) for three tuples r_1, r_2 and r_3 having fuzzy values A_{r1}, A_{r2} and A_{r3}, respectively is shown in Fig. 5.16. We can see that tuple r_2 should meet the query condition with a higher degree than r_3.

Table 5.7 Tuples from database (Fig. 5.9), which meet fuzzy query condition: *pollution = 95*

Municipality	Pollution (mg)	Matching degree
mun 3	85; 95; 100; 110—trapezoidal number	1
mun 2	90; 100; 105—triangular number	0.5

Fig. 5.16 Query condition WHERE A $< q$, in which A is a fuzzy attribute

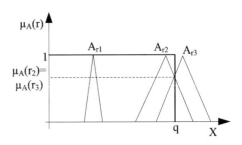

Fig. 5.17 Triangular fuzzy number in crisp condition $< q$

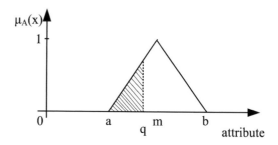

The solution can be reached by the proportion of fuzzy data surface covered by crisp query and the whole surface of fuzzy data. This proportion can be calculated by integrals [19] or cardinalities [28]. Concerning the former, membership degree for the operator $<$ can be calculated as:

$$\mu(r) = \frac{\int_a^q \mu_A(x)dx}{\int_a^b \mu_A(x)dx} \qquad (5.9)$$

where q is the value of crisp condition, and values a and b delimit the support of fuzzy data for tuple r. Similarly, membership degree for the operator $>$ can be calculated as:

$$\mu(r) = \frac{\int_q^b \mu_A(x)dx}{\int_a^b \mu_A(x)dx} \qquad (5.10)$$

where parameters have the same meaning as in (5.9).

Example 5.13 User would like to select municipalities which have pollution smaller than 38 mg.

The procedure is as follows: Check, whether the attribute pollution is fuzzy. The answer is in the IS_FUZZY table (Fig. 5.9). If attribute is fuzzy, then activate calculation of crisp condition on fuzzy values, i.e. select municipalities which meet the condition $m < q$ for singletons and $a < q$ for non-singletons.

Finally, calculate matching degree of each selected municipality. For non-singletons, calculation is divided into two cases: if $b < q$, the matching degree is 1, otherwise calculation of the surface by (5.9) is activated.

From the fuzzy meta model shown in Fig. 5.9 only municipality $mun1$ is selected. For this tuple $38 \in [35, 45]$ and therefore calculation of surface is activated. The numerator of (5.9) is calculated as:

$$\int_{35}^{38} \frac{x - 35}{5} dx = 0.9 \qquad (5.11)$$

Similarly, the value of denominator is calculated. Finally, the matching degree is 0.18. If the condition were pollution >38, the satisfaction degree would be 0.82. □

The importance of bounded support is evident. If we express pollution as Gaussian function, then belonging to query cannot be properly expressed, because the support of this fuzzy set is unlimited and therefore shape's surface cannot be calculated.

5.6.2.2 Fuzzy Queries on Fuzzy Attributes

Variety of comparison operators is applicable. In this section focus is on cardinality, similarity, possibility and necessity.

Flexible conditions by cardinality, possibility and similarity

In Chap. 2 fuzzy queries of the condition, e.g. *attribute A is small* are realized on crisp relational databases. In this section the same kinds of queries are focused on querying fuzzy attributes. In this kind of queries, where values of attribute *A* are fuzzy, the possibility measure is applied. By this measure the possibility that the tuple *r* from a database belongs to the fuzzy concept is computed.

Example 5.14 An user wants to see the low polluted municipalities by the query condition

WHERE pollution is small

Analogously to Sect. 2.3, the query is divided into two steps: selecting relevant tuples and calculating their matching degrees. If the support of fuzzy value *A* touches the support of the fuzzy condition *F*, i.e. supp(A) $\bigcap$ (F) $\neq \varnothing$, then the tuple partially or fully meets the condition. In other words, the intersection (1.30) means that satisfaction degree is greater than 0. The graphical interpretation of satisfaction degree calculations is in Fig. 5.18 and the solution is in Table 5.8. □

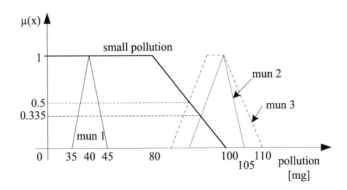

Fig. 5.18 Possibility that fuzzy value belongs to concept *small pollution*

Table 5.8 Low polluted municipalities

Name	Pollution (mg)	Matching degree
mun 1	35; 40; 45—triangular number	1.000
mun 3	90; 100; 105—triangular number	0.500
mun 2	85; 95; 100; 110—trapezoidal number	0.335

The next option is the generalization of the = operator by fuzzy similarity relation [29]. Let A be fuzzy set of attribute and B fuzzy set expressing query condition, then fuzzy similarity is expressed as:

$$F_s(A, B) = t((A \subseteq_F B), (B \subseteq_F A)) \tag{5.12}$$

where index F stands for fuzzy inclusion and t is a t-norm. If minimum function is used as t-norm, then the similarity is expressed as [29]:

$$F_s(A, B) = \frac{card(A \cap B)}{\max(card(A), card(B))} \tag{5.13}$$

where the cardinality of fuzzy set is expressed either by Eq. (1.7), if membership function is assigned to each element, or by area of membership function (1.21). The intersection (1.30) is mainly subnormalized fuzzy set, i.e. height is lower than 1. In this case, the cardinality is calculated by multiplying obtained area with the height of intersection.

The possibility function can be used in all examined cases (is small, is medium, is high, is about s), whereas similarity can be used only for the condition (is about s). Let us demonstrate this issue in the following example.

Example 5.15 Let us examine two query conditions: *pollution is small; pollution is about 41 mg.*

The first query condition is expressed by L fuzzy set with parameters $m = 80$ and $b = 100$. For simplicity, municipalities *mun2* and *mun3* are not considered. Query condition and fuzzy value for *mun*1 are depicted in Fig. 5.18. Applying possibility function we get the satisfaction degree of 1 which is a fully expected solution. Let us try calculation by cardinalities:

$card(A \cap B) = \frac{5+5}{2} = 5$, $card(A) = \frac{5+5}{2} = 5$, $card(B) = (80 - 0) + \frac{100-80}{2}$
$= 90$, $F_s(A, B) = \frac{5}{90} = 0.055$.

Apparently, the solution is far from the right one.

The second query condition is expressed by a non-symmetric triangular fuzzy set with parameters $a = 35$, $m = 41$ and $b = 45$. The same municipality is considered again. Query condition as well as fuzzy value for examined municipality are depicted in Fig. 5.19. The possibility function returns value of 0.908, because the maximum

Fig. 5.19 Possibility and
similarity that fuzzy value is
equal to concept *pollution*
about 41 mg

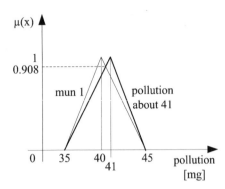

of intersection is reached for pollution of 40.45 mg. The result by cardinality is as
follows:

$$card(A \cap B) = 0.908 \cdot \frac{(40.45 - 35) + (45 - 40.45)}{2} = 4.54, \ card(A) = card(B) =$$

5, $F_s(A, B) = \frac{4.54}{5} = 0.908$.

The results are equal, which is not surprising, because these two compared sets
are almost equal. In the first condition support and core of fuzzy value is only a small
subset of the fuzzy concept. It means that fuzzy value is inside the small pollution,
but not that this fuzzy value is similar to the fuzzy concept.   □

Possibility measure can be applied for query conditions *is small, is medium, is
high*, whereas similarity by cardinalities of fuzzy sets is suitable for *more or less
equal to* comparison operators.

Fuzzy comparators

In addition to the classical comparison operators ($>$, $\gg$, $<$, $\ll$, $=$, $<=$ ...), fuzzy
conditions may contain fuzzy comparators. Each classical comparator is fuzzified
into two fuzzy comparison operators: *possibly* and *necessarily*. In this way operator
$=$ is fuzzified into operators: *possibly fuzzy equal to* and *necessarily fuzzy equal to*;
operator $>$ is fuzzified into: *possibly fuzzy greater than* and *necessarily fuzzy greater
than* and so on. All these operators are examined in [16]. In this section fuzzified
equal to and *less than* comparators are illustrated.

Possibly fuzzy equal to (PFE)

The query condition may be expressed as: WHERE attribute *A PFE F*, in which *F*
stands for fuzzy number and attribute *A* contains fuzzy data. The query matching
degree for tuple *r* is calculated by possibility measure (1.28) as:

$$\mu^r_{A_{PFE}F} = Poss(F, A) = \sup_{x \in X}[\min(\mu_A(x), \mu_F(x))] \tag{5.14}$$

where *x* stands for all possible values in the support of fuzzy data *A* for tuple *r*.
When crisp comparator is used and *F* is a singleton, we obtain the same solution as
in Example 5.12.

Necessarily fuzzy equal to (NFE)
Matching degree for tuple r to the query condition: WHERE attribute A *NFE F* is computed by necessity measure:

$$\mu^r_{A_{NFE}F} = Nec(F, A) = \inf_{x \in X}[\max(\mu_F(x), 1 - \mu_A(x))] \qquad (5.15)$$

Example 5.16 An architect is interested to find buildings the age of which is possibly equal to 220 years, where term *equal to 220* is expressed as triangular fuzzy set with parameters $a = 200$, $m = 220$ and $b = 240$. Relation BUILDING is a part of database shown in Fig. 5.9. It is clear that buildings $b1$, $b2$ and $b4$ do not match. Matching degree for building $b3$ is 0.429, because supremum of intersection between attribute's value and condition is in age 208.57. The answer is graphically illustrated in Fig. 5.20. □

In order to avoid unnecessary calculations (membership degrees for tuples which neither fully nor partially match the condition), SQL query may be applied. The condition for triangular and trapezoidal fuzzy numbers in example above should select tuples in which value b is greater than 200 and value a is less than 240; for singleton fuzzy number value m should be found in the [200, 240] interval.

Possibly fuzzy less than (PFL)
Matching degree for tuple r to the condition: WHERE attribute A *PFL F* is calculated by the modified possibility measure, where F is expressed as L type fuzzy set, and attribute A is trapezoidal fuzzy number [16], as:

$$\mu_{A_{PFL}F}(x)^r = \begin{cases} 1 & \text{for } m_{1A} \leq m_F \\ \frac{a_A - b_F}{(m_F - b_F) - (m_{1A} - a_A)} & \text{for } m_{1A} > m_F \wedge a_A < b_F \\ 0 & \text{for } a_A \geq b_F \end{cases} \qquad (5.16)$$

where a_A, m_{1A}, m_{2A}, b_A stand for parameters of trapezoidal fuzzy number A (Fig. 1.7) and a_F and m_F for L type fuzzy set F (Fig. 1.8). When A is a triangular fuzzy number, then its modal value m corresponds with m_1.

Fig. 5.20 Answer to the query *age is possibly equal to 220* years for building $b2$

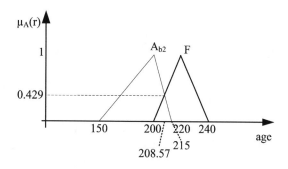

Necessarily fuzzy less than (NFL)

Matching degree for tuple r to the condition: WHERE attribute A *NFL* F, where F is expressed as L type fuzzy set, and attribute A is trapezoidal fuzzy number is calculated as [16]:

$$\mu_{A_{NFL}F}(x)^r = \begin{cases} 1 & \text{for } b_A \leq m_F \\ \frac{m_{2A} - b_F}{(m_F - b_F) - (b_A - m_{2A})} & \text{for } b_A > m_F \wedge m_{2A} < b_F \\ 0 & \text{for } m_{2A} \geq b_F \end{cases} \tag{5.17}$$

where parameters have the same meaning as in (5.16). When fuzzy data A is expressed by triangular fuzzy number, then its modal value m corresponds with m_2.

Example 5.17 Let us have in a database three tuples having fuzzy numbers for attribute A expressed by notation (1.33) as $A_{r1}(1, 2.5, 4)$, $A_{r2}(4, 5, 6)$ and $A_{r3}(8, 10, 12)$. Two query conditions are Q_{f1}: *attribute A is possibly fuzzy less than F*, and Q_{f2}: *attribute A is necessarily fuzzy less than F*, where F is expressed as L type fuzzy set with parameters $m = 3$ and $b = 9$. Tuples and query condition are plotted in Fig. 5.21. Answer to query Q_{f1} is calculated by (5.16) as: $A_{Q_{f1}} = \{(r1, 1), (r2, 0.71), (r3, 0.125)\}$. In the same way, answer to query Q_{f2} is calculated by (5.17) as: $A_{Q_{f2}} = \{(r1, 0.87), (r2, 0.57), (r3, 0)\}$. Matching degrees calculated by necessity functions are lower or equal than matching degrees calculated by possibility functions.

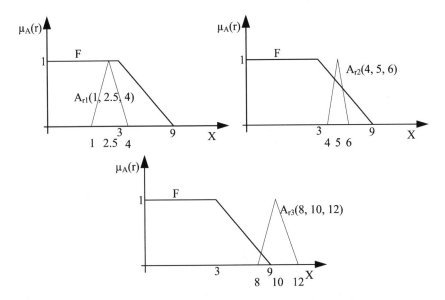

Fig. 5.21 Three tuples and fuzzy condition for calculation of membership degree to conditions *fuzzy possibly less than* and *fuzzy necessarily less than*

In order to avoid unnecessary calculations (membership degrees for tuples which neither fully nor partially match the condition), SQL query should select only tuples with degrees greater than 0. Hence, SQL query selects tuples which have value a of considered attribute lower than value b of fuzzy condition, when the *possibly less than* comparator is used, or selects tuples which have value m of considered attribute lower than value b of fuzzy condition, when the *necessarily less than* comparison operator is used. □

In the same way, other fuzzy comparators can be applied as well. Calculating membership degrees for all tuples is an unnecessary computational burden, because usually a (small) subset of tuples meets the query condition. Hence, in the first step we should select only tuples, which have matching degree greater than 0. For fuzzy queries in traditional databases this option is illustrated on procedure in Sect. 2.3. In the same manner, that procedure may be extended to fuzzy database. In the second step calculation of membership degrees is realized.

The query conditions can be further created by aggregation functions, e.g. *select districts where average of pollution in municipalities is possibly fuzzy equal to 95 mg of the pollutant Pl* or *select municipalities where age of buildings is necessarily fuzzy greater than 50 years and size is <250 and flooded level is small.* The combination of querying possibilities is literally unlimited. It is impossible to discuss all of them. Our intent is to discuss the main concepts and the user can further combine them.

5.7 Linguistic Summaries on Fuzzy Data

LSs focused on crisp data is the topic of Chap. 3. When fuzzy data are stored in a fuzzy relational database, then paradigm created for crisp data can be used. In this section, similarly as in Chap. 3, the focus is on classical protoforms.

The difference in summarizing from crisp and fuzzy data is in calculation of proportions consisting of summarizer and restriction. It is not possible to apply (3.2) and (3.8), because those calculations are focused on crisp data. Regarding the computation of linguistic quantifiers, we can use the same structures and procedures, which were demonstrated in Sect. 3.3.

The procedure for calculating membership degrees inside proportions is extended to cover fuzziness in data and comparison operators.

The first extension consists of crisp comparators on fuzzy data to cover LSs such as *most of buildings have age = 200* or *most of buildings of age = 200 has size < 120* m^2, where age is fuzzy attribute. Membership degrees are computed by possibility measure, cardinalities or proportions in the same way as was illustrated in Sect. 5.6.2.1.

Table 5.9 Attributes age and size of buildings

Building	Age (years)	Size (m^2)
b1	(190, 200, 210, 220)	110
b2	(160, 190, 210)	122
b3	(50)	128
b4	(195, 205, 220)	210
b5	(200)	120
b6	(190, 205, 210, 230)	95
b7	(180, 190, 230)	128
b8	(200)	110
b9	(190, 200, 210)	90
b10	(203)	88

Example 5.18 An architect, who examines the size of older buildings, wants to find out, whether most of buildings with age equal to 200 years have small size. The summarizer is in the same structure as for summarizers in Chap. 3, because size is crisp attribute in a database. The restriction part age of building is fuzzy attribute stored in a fuzzy meta model. For the sake of simplicity, attributes age and size are shown in Table 5.9.

Condition *age = 200* is converted into fuzzy singleton ($m = 200$). In this way possibility measure (1.28) is applied in the same manner as in Example 5.12. Term *small size* is expressed as L fuzzy set with parameters $m = 120$ and $b = 130$. The computed proportion in this summary is:

$$\frac{\sum_{i=1}^{n} t(\mu_S(x_i), \mu_R(x_i))}{\sum_{i=1}^{n} \mu_R(x_i)} = \frac{1+0.5+0+0+1+0.67+0.2+1+1+0}{1+0.5+0+0.5+1+0.67+0.75+1+1+0} = 0.83641$$

Applying quantifier *most of* (3.7) with parameters $m = 0.5$ and $n = 0.85$, the validity is 0.96.   □

The second extension is more complex. It covers fuzzy data, fuzzy comparators and fuzzy predicates. Each fuzzy comparison operator can be expressed by possibility and necessity functions. Examples of such summaries are: *most of possibly old buildings have possibly high gas consumption* and *most of possibly low polluted historic villages has necessarily high number of visitors*.

Hence, the possibilities for constructing LSs are high. Calculation of validities requires higher computational effort. On the other hand, flexibility covers large scale of problems, in which imprecision in data must not be neglected. A single LS can be consisted of linguistic quantifier, fuzzy and crisp data, fuzzy and crisp comparison operators, and fuzzy and crisp predicates.

Concerning attributes which are expressed by limited number of terms, the possibility measure is adjusted to the countable fuzzy set in the following way [17]:

$$Poss(\mu_{FD}(x), \mu_C(x)) = \max_{x \in X}[\min(\mu_{FD_i}(x), \mu_{C_i}(x))] \qquad (5.18)$$

Table 5.10 Customers age and fuzzy opinions related to solving complaints

Name	Opinion	Age
cust 1	M: 0.4; H:0.6	44
cust 2	M: 0.5; H:0.5	37
cust 3	M: 0.5; H:0.5	51
cust 4	H: 0.8; VH:0.2	34
cust 5	L:1	58

where index FD stands for fuzzy data (e.g. opinion about solving complaints in Example 5.5 of customer x), C for examined concept (e.g. positive opinion in the same example) and i for i-th linguistic term.

Example 5.19 A company is interested to examine customers' opinions from the point of view of their age. A database capable to store these data is mentioned in Example 5.5. The stored data are shown in Table 5.10. The task is to find relation between positive opinion and middle age. Hence, the LSs are of structure:

ls_1: *most of customers which have positive opinion are middle aged*;
ls_2: *about half of customers which have positive opinion are middle aged*;
ls_3: *few of customers which have positive opinion are middle aged*.

The concept *positive opinion* is expressed as fuzzy set $PO = \{$(high, 0.75), (very high, 1)$\}$. The concept *middle aged* (MA) is expressed as trapezoidal fuzzy set with parameters $a = 35$, $m_1 = 40$, $m_2 = 50$ and $b = 55$.

Regarding the opinion attribute, the possibility measure is calculated as:

$$Poss(\mu_{OC}(x), \mu_{PO}(x)) = \max_{x \in X}[\min(\mu_{OC_i}(x), \mu_{PO_i}(x))]$$

where indexes OC and PO stand for opinion of a customer and concept of positive opinion, respectively.

Applying possibility measures, degrees of 0.4 for the first attribute and 1 for the second attribute for the *cust* 1, are calculated. In the same way, other membership degrees are calculated and shown in Table 5.11.

Consequently, the proportion of respondents for all three summaries is 0.6383. Applying quantifiers *most of*, *about half* and *few* with the same parameters as in

Table 5.11 Membership degrees to positive opinion and middle age

Name	μ_{PO}	μ_{MA}
cust 1	0.6	1
cust 2	0.5	0.4
cust 3	0.5	0.8
cust 4	0.75	0
cust 5	0	0

Fig. 3.4, the validities of summaries are the following: $v(ls_1) = 0.1915$, $v(ls_2) = 0.8085$ and $v(ls_3) = 0$. Finally, these LSs can be evaluated from the quality perspective. Values for measures of coverage (3.12), novelty-outlier (3.14) and simplicity (3.15) are the same for all three LSs, because they use the same proportion. More precisely, the coverage index i_c (3.11) gets value of 0.22 which implies that the coverage is 1. Measures of non-outlier and simplicity get both the value of 1. Hence, the summary ls_2 can be considered as a summary explaining stored data about customers.   □

The LS concept easily ameliorates efficiency of mining quantified textual description of either crisp or fuzzy database.

5.8   Final Remarks

Large amount of data is the subject of interest of businesses, research institutes and governmental agencies. Let us speculate upon this issue from the perspective of collecting statistical data. It is obvious that the same or similar issues can be found in many other fields. Respondents are reluctant to cooperate in official surveys what causes increase of costs for measuring data quality and for imputations. On the other hand, many (big) data are already collected, though in different structures. They can be adjusted to use in the official statistics data collections, e.g. [1, 23, 32]. Many data are of crisp nature, such as: number of sold items, number of registered tourists and the like. But many other cannot be always expressed as sharp values.

For instance, analysing and classifying sentiments from social networks [39] or from open-ended questions in surveys [3] about development of society, is another example, where variety of data is collected, but data cope with imprecision. Businesses are interested in mining patterns from their own data and data explaining many aspects of society in order to adjust marketing strategies, prepare new products, etc. Traditionally, collected data are stored as crisp causing that the fuzziness of real world is lost. Now we have possibility to record this fuzziness by deciding which data should be stored as fuzzy.

Illustrative examples used in this chapter are related to municipal statistics and customers. Many other data (crisp and fuzzy) can be related to territorial units, even in short time intervals, such as traffic flow or energy consumption. Furthermore, opinions from social networks and questionnaires provide additional valuable information for business and policy decision making.

Summing up, interest in data increases. We should not neglect fuzzy nature of the data. Therefore, the fuzzy database option should be kept on the shelf and used, when it is not possible to keep all data as crisp values. In addition, approaches, such as fuzzy meta model, allow keeping fuzzy values for selected attributes, whereas all other data are kept as crisp ones in usual way in relational databases.

References

1. Altin, L., Tiru, M., Saluveer, E., Puura, A.:Using passive mobile positioning data in tourism and population statistics. In: New Techniques and Technologies for Statistics (NTTS 2015), Brussels (2015)
2. Aguilera, A., Mata-Toledo, R., Subero, A., Monger, M., Gupta, P.: On an extension of fuzzy aggregate functions for databases. J. Inf. Syst. Oper. Manage. **7**, 1–8 (2013)
3. Balbi, S., Triunfo, N.: Statistical tools in the joint analysis of closed and open-ended questions. In: Davino, C., Fabbris, L. (eds.) Survey Data Collection and Integration, pp. 61–72. Springer, Berlin Heidelberg (2012)
4. Bojadziev, G., Bojadziev, M.: Fuzzy logic for business, finance and management, 2nd edn. World Scientific Publishing Co., Ltd., London (2007)
5. Borkowski, L. (ed.), Lukasiewicz, J.: Selected Works. North-Holland Publishing Company, Amsterdam and Warsaw (1970)
6. Buckles, B.P., Petry, F.E.: Extending the fuzzy database with fuzzy numbers. Inf. Sci. **34**, 45–55 (1984)
7. Buckles, B.P., Petry, F.E.: Fuzzy databases and their applications. In: Gupta, M., Sanchez, E. (eds.) Fuzzy Information and Decision Processes, vol. 2, pp. 361–371. North-Holland, New York (1982)
8. de Caluwe, R., de Tré G. (eds.): Special issue on advances in fuzzy database technology. Int. J. Intell. Syst. **22** (2007)
9. Chen, G.Q., Kerre, E.E., Vandenbulcke, J.: Normalization based on functional dependency in a fuzzy relational data model. Inf. Syst. **21**, 299–310 (1996)
10. Chen, P.P-S.: The entity-relationship model—towards a unified view of data. ACM Trans. Database Syst. **1**, 9–36 (1976)
11. Codd, E.F.: Missing information (applicable and inapplicable) in relational databases. ACM SIGMOD Rec. **15**(4) (1986)
12. Codd, E.F.: Extending the database relational model to capture more meaning. ACM Trans. Database Syst. **4**, 397–434 (1979)
13. Date, C.J.: Date on Databases: Writings 2000–2006. Apress, New York (2006)
14. Galindo, J., Medina, J.M., Aranda-Garrido, M.C.: Fuzzy division in fuzzy relational databases: an approach. Fuzzy Sets Syst. **121**, 471–490 (2001)
15. Galindo, J., Medina, J.M., Pons, O., Cubero, J.C.: A server for fuzzy SQL Queries. In: Andreasen, T., Christiansen, H., Larsen, H.L. (eds.) Flexible Query Answering Systems, vol. 1495, pp. 164–174. Springer, Heidelberg (1998)
16. Galindo, J., Urrutia, A., Piattini, M.: Fuzzy Databases—Modeling. Design and Implementation. Idea Group Publishing, Hershey (2006)
17. Hudec, M.: Storing and analysing fuzzy data from surveys by relational databases and fuzzy logic approaches. In: XXVth IEEE International Conference on Information, Communication and Automation Technologies (ICIT2015), pp. 220–225. Sarajevo (2015)
18. Hudec, M., Praženka, D.: Collecting, storing and managing fuzzy data in statistical relational databases. Stat. J. IAOS, **32**(2), 245–255 (2016)
19. Hudec, M.: Fuzzy data in traditional relational databases. In: 12th Symposium on Neural Network Applications in Electrical Engineering, pp. 195–200. Belgrade (2014)
20. Kunii, T.L.: Data plan: an interface generator for database semantics. Inf. Sci. **10**, 279–298 (1976)
21. Likert, R.: A technique for te measurement of attitudes. Arch. Psychol. **22**(140), 1–55 (1932)
22. Li, Q.: A novel Likert scale based on fuzzy set theory. Expert Syst. Appl. **40**, 1609–1618 (2013)
23. Ma, Y., van Dalen, J., De Blois, C., Kroon, L.G.: Estimation of dynamic traffic densities for official statistics. J. Transport. Res. Board **2256**, 104–111 (2011)
24. Ma, Z.M., Yan, L.: A literature overview of fuzzy database models. J. Inf. Sci. Eng. **24**, 189–202 (2008)

25. Majumdar, A.K., Raju, K.N.: GEFRED: fuzzy functional dependencies and lossless join decomposition of fuzzy relational database system. ACM Trans. Database Syst. **13**, 129–166 (1993)
26. Medina, J.M.: Pons., O., Villa, M.A.: GEFRED: a generalized model of fuzzy relational databases. J. Inf. Sci. Eng. **76**(1–2), 87–109 (1994)
27. Pavese, F.: Why should correction values be better known than the measurand true value. J. Phys. Conf. Series **459** (2013)
28. Perović, A., Takači, A., Škrbić, S.: Formalising PFSQL queries using LP1/2 fuzzy logic. Math. Struct. Comput. Sci. **22**, 533–547 (2012)
29. Perović, A., Takači, A., Škrbić, S.: Towards the formalization of fuzzy relational database queries. Acta Polytechnica Hung. **6**, 185–193 (2009)
30. Petry, F.E.: Fuzzy Databases—Principles and Applications. Kluwer, Boston (1996)
31. Prade, H., Testemale, C.: Representation of soft constraints and fuzzy attribute values by means of possibility distributions in databases. In: Bezdek, J. (ed.) Analysis of Fuzzy Information (Vol, II): Artificial Intelligence and Decision Systems, pp. 213–229. CRC Press (1987)
32. Puts, M., Daas, P., Tennekes, M.: High frequency road sensor data for official statisitics. In: New Techniques and Technologies for Statistics (NTTS 2015), Brussels (2015)
33. Shenoi, S., Melton, A.: Functional dependencies and normal forms in the fuzzy relational database model. Inf. Sci. **60**, 1–28 (1992)
34. Škrbić, S.: Using fuzzy logic in relational databases. Ph.D. Dissertation. University of Novi Sad, Novi Sad (2008)
35. Škrbić, S., Racković, M.: PFSQL: a fuzzy SQL language with priorities. In: 4th International Conference on Engineering Technologies, pp. 58–63, Novi Sad (2009)
36. Škrbić, S., Racković, M., Takači, A.: Towards the methodology for development of fuzzy relational database applications. Comp. Sci. Inf. Syst. **8**, 27–40 (2011)
37. Sözat, M.I., Yazici, A.: A complete axiomatization for fuzzy functional and multivalued dependencies in fuzzy database relations. Fuzzy Sets Syst. **117**, 161–181 (2001)
38. Takači, A., Škrbić, S.: Data model of FRDB with different data types and PFSQL. In: Galindo, J. (ed.) Handbook of Research on Fuzzy Information Processing in Databases, pp. 407–434. Information Science Reference, Hershey (2008)
39. Torres van Grinsven, V., Snijkers, G.: Sentiments and perceptions of business respondents on social media: an exploratory analysis. J. Official Stat. **31**, 283–304 (2015)
40. Umano, M., Fukami, S.: Fuzzy relational algebra for possibility-distribution-fuzzy-relation model of fuzzy data. J. Intell. Inf. Syst. **3**, 7–27 (1994)
41. Urrutia, A., Tineo, L., Gonzales, C.: FSQL and SQLf: Towards a standard in fuzzy databases. In: Galindo, J. (ed.) Handbook of Research on Fuzzy Information Processing in Databases, pp. 270–298. Information Science Reference, Hershey (2008)
42. Viertl, R.: Fuzzy data and information systems. In: 15th International WSEAS Conference on Systems, pp. 83–85, Corfu (2011)
43. Vucetic, M., Hudec, M., Vujošević, M.: A new method for computing fuzzy functional dependencies in relational database systems. Expert Syst. Appl. **40**, 2738–2745 (2013)
44. Yang, M.F., Chou, Y.T., Lo, M.C., Tseng, W.C.: Fuzzy time distribution in PERT model. In: International Multiconference of Engineers and Computer Scientists (IMECS 2014), pp. 1084–1087, Hong Kong (2014)
45. Zadeh, L.A.: Fuzzy sets as a basis for a theory of possibility. Fuzzy Sets Syst. **1**, 3–28 (1978)
46. Zadeh, L.A.: Similarity relations and fuzzy ordering. Inf. Sci. **3**, 177–200 (1971)
47. Zemankova-Leech, M., Kandel, A.: Implementing imprecision in information systems. Inf. Sci. **37**(1–3), 107–141 (1985)

Chapter 6
Perspectives, Synergies and Conclusion

Abstract Fuzziness can be found in many areas of daily life. Hence, fuzziness cannot be always expressed with one aspect and solved by one approach. It implies that different approaches should cooperate. In addition, many tasks, for example, in smaller businesses are not extremely demanding for complex tools, but rather they look for overviews of problems from different aspects. This short concluding chapter is focused on cooperation between fuzzy queries, summaries and inferences with respect to fuzzy and crisp data.

6.1 Perspectives

The conclusion (decision, action, advice, etc.) is reliable, only when sufficient number of relevant data is at disposal. Nowadays, the data masses are growing so fast, that they supersede the human capability of perception to recognize relevant data, as well as to detect relations and dependences among attributes to make the conclusion. In order to cope with this problem robust tools and approaches are inevitable.

Companies and other institutions may have numerous data sources to support their decisions. However, researches have shown that many business decisions are based on intuition, heuristics and impressions [3–5], although the potential in data is sound. Presumably, one of reasons is that traditional dichotomous approaches and tools cannot provide environment for incorporating subjectivity and imprecision of real world [6]. According to Lim [1], data-driven competitive predictions have more predictive accuracy than predictions based on informal intelligence. We can say that the same holds in tasks, where institutions wish to mine relevant knowledge from their internal data (customers, production, etc.) supported by external data explaining various aspects of society (inflation, average wages, municipal statistics, censuses and the like).

Speculating about these sentences leads us to merging data, informal intelligence (experiences, imprecision, subjectivity) and formal mathematical procedures. In order to gain benefits from data and incorporate subjectivity and experiences of workers, the less complex and robust fuzzy tool(s), capable to cope with fuzziness of the real world, is an option which should be considered.

© Springer International Publishing Switzerland 2016
M. Hudec, *Fuzziness in Information Systems*,
DOI 10.1007/978-3-319-42518-4_6

In flexible queries (Chaps. 2 and 5), summarization (Chaps. 3 and 5) and inference (Chap. 4), the first step is construction of fuzzy sets for linguistic terms. The simplest way is to give users freedom to directly define parameters of fuzzy sets. Another way is based on mining fuzzy sets parameters from the current database content. Actually, merging these ways creates more sophisticated fuzzification. In the next steps all three examined approaches do not need to cooperate, while solving tasks using data from crisp or fuzzy databases. However, their cooperation can be beneficial for users.

Before to proceed, we should emphasize that, although fuzzy logic has considerable potential for solving variety of problems, it is not a medicine for all problems related to uncertainties. According to [7], *fuzzy epoch* has already begun. Nowadays, information systems and databases are usual. The same holds for fuzzy controllers. It can be expected that fuzzy logic tools will be usual in businesses and public institutions. A significant advantage of fuzzy logic and fuzzy set theory is its generality, enabling us to adapt existing approaches and create new ones for solving problems, which will emerge in the future. Since the first paper in fuzzy set theory has been written, a large variety of theoretical contributions and applications appeared: from controlling technical systems to support business intelligence.

6.2   Synergy

This part is focused on several possibilities for integrating approaches examined in previous chapters. Many other possible approaches are not discussed, due to space limitation of the textbook and because, on our opinion, the selected approaches can solve variety of everyday tasks in businesses and public institutions. If these approaches work in a complementing, rather than competitive way, users gain further benefits. The integration is schematically plotted in Fig. 6.1. Influences and relations among these approaches are on the use case diagram shown in Fig. 6.2. This reflection is based on the argument that in fuzzy logic it is easy to layer on more functionality without starting again from scratch. This integration is a valuable support in business informatics. It integrates approaches suitable for managing fuzziness of the real world, which may influence companies' results.

6.2.1   Fuzzy Inference and Fuzzy Databases

One of disadvantages of crisp inference is the static structure of the rule base. Rules should be able to adjust to the changing environments. Although fuzzy rules seem to be static, they are adjustable to changes. Linguistic terms in rules, such as *small*, *medium* and *high*, remain the same, keeping the meaning and purpose of the rule base. Parameters of fuzzy sets can be adjusted for a particular task and stored in

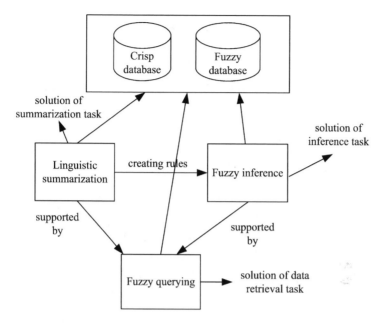

Fig. 6.1 A possible integration of querying, summarizing and inferencing from crisp and fuzzy databases

a fuzzy database. In this way, fuzzy rule base can be managed in the same fuzzy database, where fuzzy data measuring attributes' values are stored.

It was emphasized in Chap. 4 that fuzzy inference derives conclusions using a set of fuzzy IF-THEN rules and known facts which are crisp or fuzzy. Assume that a business company is interested in tailored motivation of its customers or a statistical institute would like to improve motivation of respondents in surveys. These tasks are long-term ones. Hence, the inference and classification should be realized repeatedly (in different time periods or in different regions). Rules may remain stable, only parameters of fuzzy sets are changeable to adjust to particular time period or area (e.g. set *high number of sold air-conditions* has different parameters for Rome and Oslo).

From these observations, the thought of using fuzzy database for storing parameters for fuzzy inference clearly emerges. This option is illustrated in Sect. 5.5.4. A relational database can be straightforwardly extended to fuzzy one. When users want to update parameters of fuzzy sets in rules, they could do it easily and efficiently by the already created procedures for managing (storing, modifying and deleting) fuzzy data.

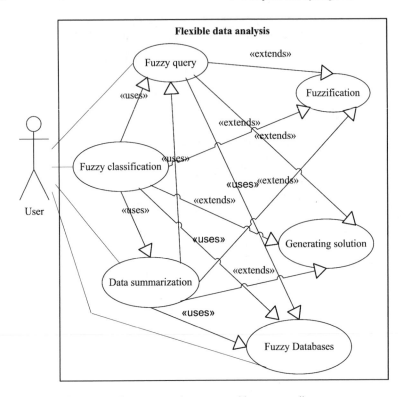

Fig. 6.2 Interactions among fuzzy approaches expressed by use case diagram

6.2.2 Linguistic Summaries and Fuzzy Inference

LSs express relational knowledge in the data. If the quality of LS is significant, then the summary can be converted into the weighted IF-THEN rule. Quality measures discussed in Sect. 3.6.1 find whether an LS is a good candidate for a rule. A rule base can be created from summaries of high quality. This option is discussed in Sect. 3.7.2. In this way, mining linguistic summaries can substitute complex tools for revealing rules.

When the consequences in IF-THEN rules are constants (Sugeno model), then inference task can be solved by fuzzy query approach extended with classification clauses. Otherwise, inference task should be solved by tools dedicated to fuzzy inference.

6.2.3 Fuzzy Queries for Fuzzy Classification and Summaries

Fuzzy queries play inevitable role in selecting relevant tuples from databases by variety of vague conditions examined in Chap. 2. But their role does not end here. Queries are able to support inference and summaries.

The former is discussed in Sect. 4.5 for classification tasks expressed by fuzzy IF-THEN rules. Usually, when solving a classification task by fuzzy inference, user should prepare the input data from database into the proper format for fuzzy inference tool and to convert result into a useful and understandable form. The solution for overcoming this difficulty can be integration of inference with fuzzy queries. Furthermore, tuples having values of attributes out of declared supports of fuzzy sets could cause problems in inference engines. In case of classification by fuzzy queries, these tuples are not selected and therefore excluded from further calculations.

If the classification or inference task is complex in number of attributes and fuzzy sets, and defuzzification is required, then an inference tool is inevitable.

In linguistic summaries, when the number of tuples and their attributes is relatively large, the computation of all required membership degrees for all summaries might take much time and might be costly [2]. Fortunately, queries and optimization techniques may reduce computational effort. Fuzzy queries reduce calculation of membership degrees only to tuples and attributes, which contribute to summaries. Moreover, queries are inevitable in providing parameters for calculating quality measures of summarized sentences.

References

1. Lim, L.K.: Mapping competitive prediction capability: construct conceptualization and performance payoffs. J. Bus. Res. **66**, 1576–1586 (2013)
2. Niewiadomski, A., Ochelska, J., Szczepaniak, P.S.: Interval-valued linguistic summaries of databases. Control Cybern. **35**, 415–443 (2006)
3. Mousavi, S., Gigerenzer, G.: Risk, uncertainty, and heuristics. J. Bus. Res. **67**, 1671–1678 (2014)
4. Persson, A., Ryals, L.: Making customer relationship decisions: analytics vs rules of thumb. J. Bus. Res. **67**, 1725–1732 (2014)
5. Torres van Grinsven, V.: Motivation in business survey response behavior. Ph.D. thesis, University of Utrecht (2015)
6. Zadeh, L.A.: Soft computing and fuzzy logic. IEEE Softw. **11**, 48–56 (1994)
7. Zimmermann, H.J.: Fuzzy Set Theory—and its Applications. Kluwer Academic Publishers, London (2001)

Appendix A
Illustrative Interfaces and Applications for Fuzzy Queries

Applications devoted to flexible queries have been suggested in the vast literature. Our focus is on the applications and interfaces which can be easily modified to meet variety of users' requirements and used as a basis for more complex tasks such as linguistic summaries.

In this appendix we explore design of fuzzy query interfaces in order to envelope flexibilities in predicates and connectives discussed in Chap. 2. This work is realized on the municipal statistics database of the Slovak Republic. Currently, the Statistical Office collects more than 800 attributes for 2925 municipalities. The majority of attributes are collected on yearly basis except those which contain stable values for a longer period such as the altitude above sea level, the year of the first written notice and the like. Therefore, working with such a database is an exciting task due to variety of data which could be very similar for some municipalities. In order to meet privacy issues related to municipal data and avoid any indication of advertisement (municipalities selected as the best in query) names of municipalities are anonymized.

A.1 Commutative Queries

In this part interfaces for queries in which the order of atomic conditions is irrelevant are examined.

Example A.1 An agency has decided to support development of agritourism. In order to fairly distribute resources, the agency has chosen to apply flexible conditions and provide resources proportionally to the matching degrees of selected municipalities. The relevant attributes are: altitude above sea level, ratio of arable land and population density. Hence, agency has constructed the following condition:

WHERE altitude is high AND ratio of arable land is high AND population density is small

© Springer International Publishing Switzerland 2016
M. Hudec, *Fuzziness in Information Systems*,
DOI 10.1007/978-3-319-42518-4

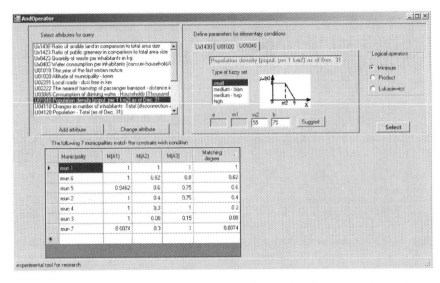

Fig. A.1 Illustrative interface for flexible commutative queries of equally relevant atomic conditions

An interface for managing such queries is shown in Fig. A.1. The available attributes are listed in the left upper part. The right upper part is designed for managing parameters of fuzzy sets. For each atomic condition there is a frame (tab control) for inserting parameters, or asking for suggestion. The functionality of the "Suggestion" button will be explained in next examples. According to the selected type of fuzzy set, only boxes for respective parameters are enabled. In Fig. A.1 the parameters for *small population density* are shown. For the aggregation, minimum t-norm (1.47) is used. The main reason is that this t-norm calculates the highest matching degree which is reflected in the proportion of obtained resources. Suppose that municipality meets all three atomic conditions with the same matching degree 0.8. If the product t-norm (1.48) is applied, then the matching degree is 0.516. The municipality will receive only a half of maximal support which is a disputable decision.

In the lower left side of interface selected municipalities ranked downwards form the best to the worst are shown. One municipality fully meets the query condition, whereas other six partially match the condition. Imagine that full amount of resources is 5 000 €. It means that the worst one, which still partially meets the condition will receive 37 €, which is a quite low amount for any activity and perhaps it is under the cost of administrative tasks related to managing support. The suitable solution is adding α-cut (1.15) to select only municipalities which significantly meet the condition.

The lower right part could be used for presenting retrieved data in a thematic map. For instance, municipalities which fully satisfy the query criterion can be marked with some colour, let us say blue, municipalities which do not satisfy the query can be marked with another colour (quite different than the one for expressing matching degree equal to 1) and municipalities which partially meet the query condition are marked again with some other colour, let us say red, having a gradient from faint hue to deep hue according to their matching degrees.   □

If the classical SQL is used and municipality *Mun 1* does not exist, the SQL would have return an empty result. Fuzzy queries mitigate the empty answer problem, but they do not entirely solve it. On the opposite side are plethoric or overabundant answers. Here we would like to point out that Łukasiewicz t-norm is one of the options for solving or at least mitigating the overabundant answers.

Flexible queries can by straightforwardly adjusted for searching similar entities. In this type of query membership functions are limited to the triangular ones (Fig. 1.5). The next example demonstrates this query.

Example A.2 In this example we are interested to find, whether municipalities with similar values of three attributes (year of the first written notice, altitude and population density) as the municipality *mun 298* exist. The interface is shown in Fig. A.2.

The query is divided into two steps. The first step consists of two parts. In the first part, user chooses municipality (in our case *mun 298*), relevant attributes and percentage of dispersion from the *m* value required for calculating the parameters *a* and *b* of triangular fuzzy sets. In the second part, SQL selects values of chosen attributes (parameters *m* for respective fuzzy sets) for the selected municipality (*mun 298*). The second step is the same as for procedure demonstrated in Example A.1 (select relevant municipalities and calculate matching degrees). The solution is shown in the lower part of interface shown in Fig. A.2.

Let us look at matching degrees of *mun p* and *mun q*. The minimum t-norm will provide different and less relevant solution than the product t-norm used in this example. Although *mun p* has lower membership degree on the third attribute (*M3*)

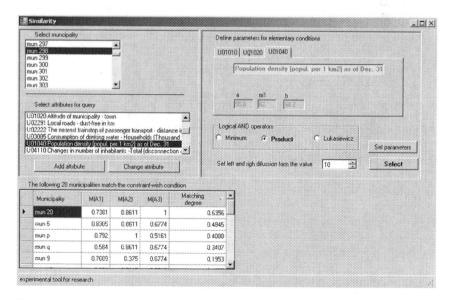

Fig. A.2 Illustrative interface for revealing similar tuples

than *mun q* on the first attribute (*M*1), the membership degrees of other two atomic conditions are a bit stronger. It implies that product t-norm is a better option. □

The similarity measure can be applied for searching municipalities which are similar to the ideal one. In this case the interface shown in Fig. A.1 can be used. The required parameters of selected attributes can be simply written into text boxes of triangular fuzzy sets. In this case, the product t-norm is the option which should be selected.

The next example illustrates query with preferences among atomic conditions. The existing interface is improved with new functionality.

Example A.3 A research institute would like to find municipalities which have high waste production per inhabitants and high water consumption in order to launch economy drive in this field. In the example, water consumption is more important than waste production. Let us say for the purpose of the example that produced waste can be recycled, but water cleaning is a more expensive task.

The interface should provide the same functionality regarding the construction of fuzzy sets and connectives as interface shown in Fig. A.1. In order to use the same interface the check box for indicating, whether atomic conditions are nonequally or equally important, is added. If this check box is marked, then user should choose weights for each atomic condition. The interface for managing queries with preferences is shown in Fig. A.3. Weight for water consumption is equal to 1 and weight for waste production is 0.5.

The solution is shown in lower part of interface (Fig. A.3). For example, *mun 4* has significantly higher matching degree in comparison to the result obtained without

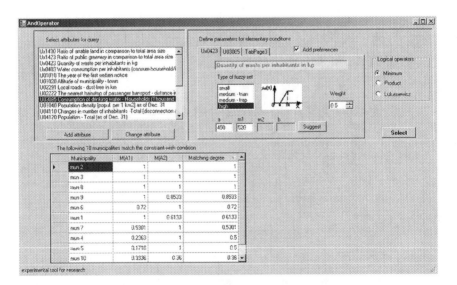

Fig. A.3 Illustrative interface for flexible commutative queries and differently relevant atomic conditions

weights. The first attribute has significantly lower importance and therefore its low membership degree does not fully influence the overall matching degree.   □

A.2   Non-Commutative Queries

In this section the focus is on the interfaces for queries in which the order of atomic conditions is relevant. The first example covers the *among* operator. The second example illustrates application of asymmetric conjunction.

Example A.4 An agency focused on local tourism activities would like to recognize municipalities with small number of beds in accommodation units among municipalities with high altitude. In order to manage such query, a suitable module and its interface are created and shown in Fig. A.4. In the left side, the list of available attributes is situated, in the same way as in aforementioned examples. Upper right side consists of two boxes: one for managing independent predicate and the other for adjusting dependent one. The user can write parameters of fuzzy sets or ask for suggestion from the application in the left box. Let user decide to express *high altitude* by R fuzzy set with parameters $a = 875$ and $m = 900$. Regarding the predicate that expresses dependent attribute, parameters of fuzzy set are generated from the subrelation delimited by the independent condition. The calculated parameters are shown in text boxes. The solution is shown in a tabular way in the lower part of the interface.   □

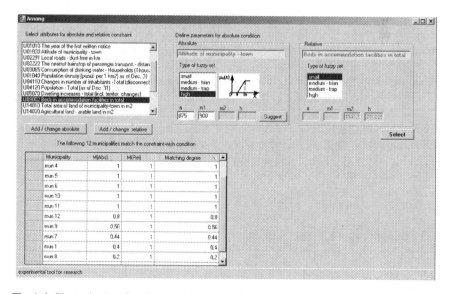

Fig. A.4 Illustrative interface for managing connective *among* between independent predicate and predicate relative to the independent one

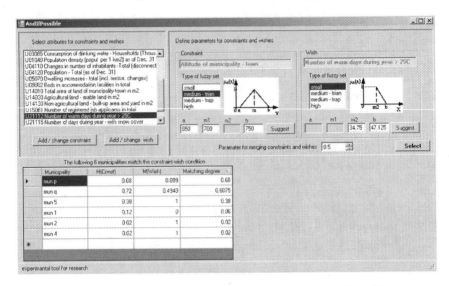

Fig. A.5 Illustrative interface for managing constraints and wishes by *and if possible* operator

Example A.5 A tourist is searching for a suitable municipality for holiday. Tourist's requirements are: altitude about 700 m and, if possible, small number of warm days. In order to manage such a query, module and its interface are created. This interface is designed in a similar way as aforementioned ones in Examples A.1, A.2 and A.4. The list of available attributes is on the left side of interface. The right side consists of two boxes: one for managing constraints and the other for wishes. Altitude about 700 m in this example means that all values higher than 650 and lower than 750 are acceptable, but with lower degree than 700. Concerning number of warm days, tourist is not aware of current content in the database and therefore asks for suggestion. Because values of this attribute are more or less uniformly distributed in the domain [0, 365] the uniform domain partition into three fuzzy sets is used. Suggested parameters are shown in the box managing wishes. It is up to user to accept or modify these parameters, if needed. Value of k (2.25) is initially set to 0.5, but the user can modify the value from 0 to 1 (from fully irrelevant wish to the equally relevant as constraint).

In the lower left side of interface, the solution appears, ranked downwards form the best municipality. The best municipality for this query is *mun p*. If for some reason it is not possible to travel there, the next one, which has a little lower matching degree, is *mun q* (Fig. A.5). ☐

A.3   Queries on Time Series

In this section fuzzy query interface is created to observe, how a given attribute changes over time.

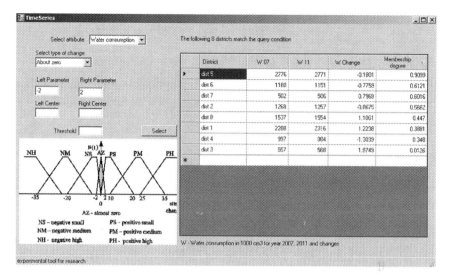

Fig. A.6 Illustrative interface for managing queries related to attribute's changes over time

Table A.1 High negative change in water consumption (high savings of water)

District	W 07	W 11	W Change	$\mu(A_{cr})$
dist 1	878	529	−39.7494	1
dist 2	1335	911	−31.7603	0.3521

Example A.6 An environmental agency wishes to have overview about changes in water consumption between the years 2007 and 2011. The first task is to find districts which have almost no change in water consumption. The second task is to find districts with measured high negative change (significantly smaller consumption). The water consumption is an aggregation-able indicator and therefore values can be aggregated from the municipality level to the district level.

Parameters for linguistic terms plotted in Fig. 2.4 are adjusted to the water consumption case, where significant changes, like e.g. in stock market exchange, are not expected. Fuzzy set *around zero* is a symmetric triangular fuzzy number (Fig. 1.5) with $m = 0$. In this example the support of this fuzzy number is limited by values $a = -2$ and $b = 2$.

The interface for managing these queries is shown in Fig. A.6. From the result we learn that eight districts partially match the query conditions. The answer for agency is that no single district with zero change exists, but there are several ones having almost zero change: *dist 1* almost meets the query condition, whereas *dist 8* very weakly meets the condition.

In the second task the *negative high consumption* requirement is expressed by L fuzzy set (Fig. 1.8) with parameters $m = -35$ and $b = 30$. The result is shown in Table A.1. From the table, agency can recognize that only one district fully matches high savings of water and another district matches it partially.   □

Appendix B
Illustrative Interfaces and Applications for Linguistic Summaries

In this part interfaces for calculating validities of linguistic summaries and their quality measures are demonstrated on the same database as in Appendix A.

Example B.1 A user wishes to know, whether most municipalities have small number of inhabitants and small total area size. In order to cope with basic LSs, an interface is shown in Fig. B.1. The left part is focused on choosing quantifier from a list of relative quantifiers and adjusting its parameters, whereas the right part is focused on constructing predicate which expresses summarizer. If more than one attribute is included, then the minimum t-norm (1.47) is used as connective. Though the interface is a bit different from the interfaces shown in Appendix A, the calculations in the application layer are based on the same procedures. Only membership degrees for municipalities, which at least partially meet both conditions in summarizer, are calculated. The parameters for fuzzy sets describing quantifier and summarizer are shown in Fig. B.1. The validity of this LS is 0.5305.

When summarizer contains only atomic predicate: small number of inhabitants, then the validity is 0.9131. Hence, an abstract: *most municipalities have small number of inhabitants* explains the whole database. Adding atomic conditions, the summarizer becomes more restrictive which implies on the validity of LS. □

Example B.2 An agency for agriculture examines relation between ratio of arable land and altitude above sea level. Agency is interested to learn, whether the majority of arable land (in comparison to total area size of municipality) is situated in municipalities having small altitude above sea level. In order to reveal the validity of this summary, the interface shown in Fig. B.2 is created. The interface is similar to the interface of basic LS (Fig. B.1). The difference is in added restriction part. The calculation is based on the (3.8). In this type of LS parameters of fuzzy sets are mined from the database and suggested to the user for approval. The user can modify them or accept them by clicking the Calculate button. Only input boxes related to selected type of fuzzy set are marked as editable. In case of high ratio of arable land, parameters a and m_1 are available, because these two parameters define R fuzzy set (1.23).

© Springer International Publishing Switzerland 2016
M. Hudec, *Fuzziness in Information Systems*,
DOI 10.1007/978-3-319-42518-4

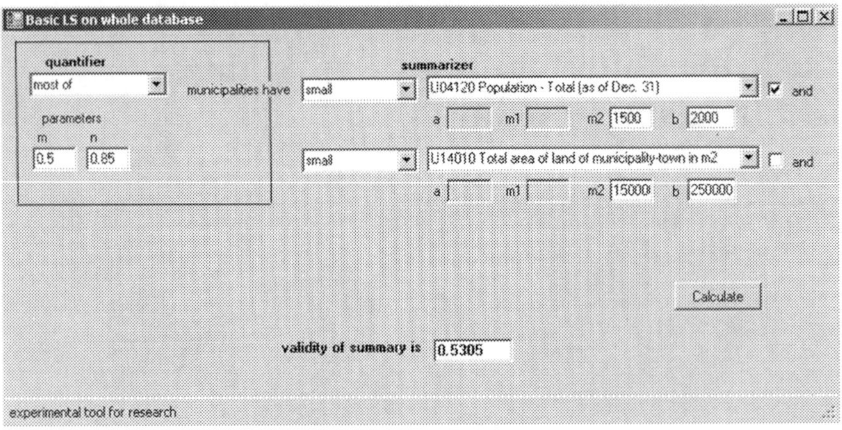

Fig. B.1 Illustrative interface for creating basic LS and computing their validities

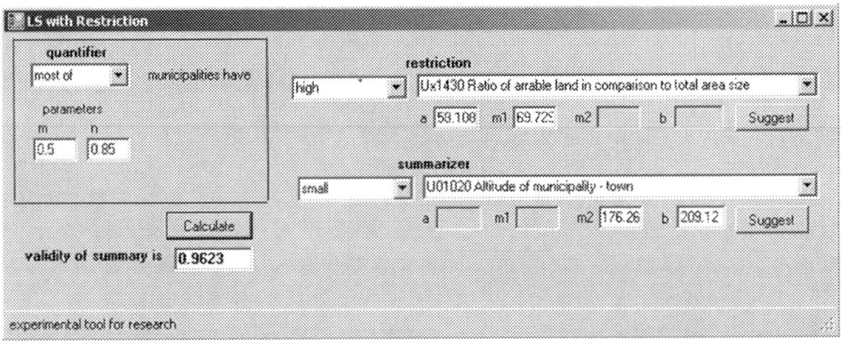

Fig. B.2 Illustrative interface for creating LSs with restriction and calculating their validities

The validity of this LS is 0.9623. Hence, an agency concludes that the sentence *most of municipalities with high ratio of arable land have small altitude above sea level* holds.   □

Interfaces for building LSs and providing their validities can be straightforwardly extended with sliders in order to display quality measures.

Example B.3 A user is interested to evaluate two LSs. The first one is of the structure *most of municipalities with high ratio of arable land have small population density*. The validity of summary is 0.814. Further, the coverage index i_c (3.11) gets value of 0.1421 for $n = 2924$ municipalities, which implies that the coverage C (3.12) gets value of 0.9927, where $r_1 = 0.02$ and $r_2 = 0.15$. Finally, the outlier measure (3.14) is 0.0047. These values guided us to conclude that this LS is of a high quality.

The interface for calculating LS and its quality is shown in Fig. B.3. This interface is an extension of the already created interface shown in Fig. B.2. In the lower part

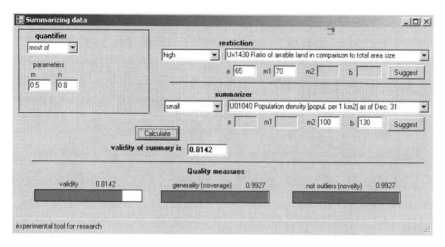

Fig. B.3 Illustrative interface for creating LSs with restriction and calculating their quality measures

of the interface three sliders graphically express quality measures (3.8), (3.12) and (3.14), respectively. In order to present uniformly informing sliders (value 1 is the best for all of them), instead of measure O (3.14), the measure $1 - O$ is displayed.

Another LS is of the structure: *most of municipalities with high number of warm days (over* 25 °C*) have small amount of produced waste per inhabitant.* The validity of summary is 1. But the coverage index gets value 0.01283. Therefore, the coverage gets value 0, when we use the same values for parameters r_1 and r_2. The outlier measure is 1 which means that this LS is based on outliers. Hence, we conclude that this LS is of low quality. Focusing only on validity might mislead us to the conclusion that this LS summarizes data in a better way than the former LS of this example. □

Index

© Springer International Publishing Switzerland 2016
M. Hudec, *Fuzziness in Information Systems*,
DOI 10.1007/978-3-319-42518-4

Printed in the United States
By Bookmasters